Industrial Ethernet

3rd edition

JOHN JAKKER

Industrial Ethernet

2nd Edition

How to Plan, Install, and Maintain TCP/IP Ethernet Networks: *The Basic Reference Guide for Automation and Process Control Engineers*

By Perry S. Marshall
and John S. Rinaldi

ISBN-10: 1-55617-892-1
ISBN-13: 978-1-55617-892-4

ISA
67 Alexander Drive
P.O. Box 12277
Research Triangle Park, NC 27709
www.isa.org

Library of Congress Cataloging-in-Publication Data
Marshall, Perry S.
  Industrial ethernet :how to plan, install, and maintain TCP/IP
ethernet networks :the basic reference guide for automation and process
control engineers / by Perry S. Marshall and John S. Rinaldi.– 2nd ed.
    p. cm.
  Includes index.
  ISBN 1-55617-892-1 (pbk.)
  1. Ethernet (Local area network system) I. Rinaldi, John S. II.
Title.
  TK5105.8.E83M369 2004
  004.6'8–dc22
                              2004019923

Notice
    The information presented in this publication is for the general education of the reader. Because neither the author nor the publisher have any control over the use of the information by the reader, both the author and the publisher disclaim any and all liability of any kind arising out of such use. The reader is expected to exercise sound professional judgment in using any of the information presented in a particular application.
    Additionally, neither the author nor the publisher have investigated or considered the affect of any patents on the ability of the reader to use any of the information in a particular application. The reader is responsible for reviewing any possible patents that may affect any particular use of the information presented.
    Any references to commercial products in the work are cited as examples only. Neither the author nor the publisher endorse any referenced commercial product. Any trademarks or tradenames referenced belong to the respective owner of the mark or name. Neither the author nor the publisher make any representation regarding the availability of any referenced commercial product at any time. The manufacturer's instructions on use of any commercial product must be followed at all times, even if in conflict with the information in this publication.

*This book is dedicated to the
Master Engineer,
whose works inspire and challenge
all designers. His creations are
beautiful, adaptable, robust,
and supremely equipped for their purpose.*

Table of Contents

About the Authors

Perry Sink Marshall is an author, speaker and consultant who advises technology companies on product definition, marketing and new customer acquisition. He specializes in control systems and communications, and in the automation industry actively works with the major journals, vendors and networking trade organizations. He has a BSEE from the University of Nebraska. You can contact him via his website, www.perrymarshall.com.

John S. Rinaldi has worked on Industrial and Building Networking for more than 20 years as both a user and a product developer. His experience includes low-level device networking, sensor bus communications, Industrial Ethernet, and SCADA systems. Current interests include remote monitoring, SOAP/XML communications, and machine-to-machine communications. His company, Real Time Automation, is a leader in Industrial and Building Automation networking software, turnkey systems, and add-on networking hardware. He can be reached through the company Web site at www.rtaautomation.com.

A special thanks to the following people who helped with the content of this book:

Deon Reynders of IDC (www.idc-online.com) is the author of "Practical TCP/IP & Ethernet Networking for Engineers & Technicians" and who also teaches IDC's excellent introductory Industrial Ethernet course for beginners. He proofread the manuscript.

Deepak Arora is a founder of Saltriver Infosystems (www.saltriver.com) and does technical support, market research and sales of networked and wireless business systems. He proofread the manuscript and provided many illustrations.

Lynn August Linse (www.linse.org) is one of the automation industry's sharpest networking application engineers and programmers, and shares my vision for the convergence of networks and protocols. He helped by suggesting content improvements.

George Karones of Contemporary Controls (www.ccontrols.com) is a hardware developer and networking application specialist. He provided important information on embedded components.

Vivek Samyal (web@tannah.net) is an automation and web applications programmer and webmaster who specializes in database work and Application Side Programming. He supplied information on network analysis tools.

Shandra Botts at ISA took on the tedious process of editing this book.

Laura, my wife, is considered by most to be a saint. She put up with me during the complex process of writing this book.

1.0—What Is Industrial Ethernet?

1.1 Introduction

Industrial Ethernet is the successful application of IEEE 802.3 standards with wiring, connectors, and hardware that meet the electrical noise, vibration, temperature, and durability requirements of factory equipment, and network protocols that provide interoperability and time-critical control of smart devices and machines.

Industrial Ethernet is a specialized, rigorous application of standard "office Ethernet" technology that adds any or all of the following requirements:

- *Mission critical:* Downtime is much less tolerable in the factory than the office. When an office network goes down, you go get a cup of coffee and check your e-mail later. When a factory goes down, you choke down your last mouthful of coffee, run into the plant, and fix the problem as fast as possible! The effects of downtime are less isolated in a manufacturing facility.

- *Harsh environment:* Factory equipment is not usually installed in air-conditioned hall closets. It's more likely to be bolted to a robotic welder or oil rig. Temperature extremes and vibration threaten garden-variety hardware, cables, and connectors. Device selection, installation, and proper wiring practices are crucial.

- *Electrical noise:* Ordinary 110 VAC circuits are not the norm in factories. Industrial Ethernet devices are often used with high-current 480 VAC power lines, reactive loads, radios, motor drives, and high-voltage switchgear. Network communication must continue reliably despite these hazards.

- *Vibration:* Industrial Ethernet "smart devices" are, by definition, mounted on machines. Machines move and shake. Velcro and "telephone connectors" may not be up to the task.

- *Powered devices:* Some devices must be powered by the network cable itself. Many automation devices operate at 24 VDC. New methods are in the works for powering these devices with Ethernet.

- *Security:* The data in your factory is not necessarily more worthy of protection than the data in your office, but the threats are different. Factory equipment is vulnerable to hackers, of course, but accidental disruptions created by yourself or your staff are much more likely. Specific precautions must be taken.

- *Legacy devices:* Real automation systems are a mix of new, nearly new, old, older, and pre-Mesozoic Era equipment from incompatible vendors. Industrial Ethernet must link serial protocols, legacy networks, and fieldbuses.

- *Interoperability:* Ethernet devices must communicate with each other, with PCs, and possibly with Internet/Web applications. The existence of an Ethernet jack is no guarantee of openness, interoperability, or compatibility. You must ask the right questions when making purchases.

- *Levels of priority:* Some machine-control information requires real-time, deterministic responses. Other data is much less urgent. It's important to recognize different priority levels for different kinds of data.

- *Performance:* Beyond physical robustness are subtle characteristics of software drivers, routers, and switches, such as hidden latencies, jitter, limited numbers of connections, and behavior under erratic conditions.

- *Connectivity to other local area networks (LANs):* Most Industrial Ethernet systems must be bridged to business intranets and the Internet. Serious problems can be introduced on both sides if this is not done with care.

- *The IT Department vs. the Automation Department:* Ethernet is precisely the place where two equally valid but conflicting views of "systems" and "data" come together. You must proceed with care to avoid a battle between company fiefdoms, all-out mutiny, or even a brand new pair of cement shoes.

- *Mastery of the basics:* No matter how good your equipment is, if you don't apply proper knowledge of Ethernet, Transmission Control Protocol/Internet Protocol (TCP/IP), and sound installation practices, your system will never work right.

Industrial Ethernet is a reference book that addresses each of these concerns and lays down the basic nuts and bolts of Ethernet and TCP/IP. After reading this book, you'll know the basics of the world's most popular network, you'll be able to plan Ethernet projects, and you'll know the right questions to ask when you talk to vendors.

Ethernet is the worldwide de facto standard for linking computers together. Ethernet connects hundreds of millions of computers and smart devices across buildings, campuses, cities, and countries. Cables and hardware are widely available and inexpensive ("dirt cheap" in the case of ordinary office-grade products), and software is written for almost every computing platform.

Ethernet is now a hot topic in automation, where industry-specific networks have dominated: Profibus, DeviceNet, Modbus, Modbus Plus, Remote I/O, Genius I/O, Data Highway Plus, Foundation Fieldbus, and numerous serial protocols over the electrical standards of EIA RS-232, RS-422, and RS-485.

In some cases, Ethernet is displacing these networks. In nearly all cases, Ethernet is being used in demanding installations alongside them. This book gives a basic understanding of Ethernet's strengths, weaknesses, fundamental design rules, and application guidelines. It addresses the unique demands of the factory environment, intelligent devices, and the most common automation applications and protocols. *Industrial Ethernet* provides basic installation and troubleshooting recommendations to help your projects work right the first time.

1.2 A Very, Very Short History of Ethernet and TCP/IP

Ethernet originated at Xerox Palo Alto Research Center (PARC) in the mid-1970s. The basic philosophy was that any station could send a message at any time, and the recipient had to acknowledge successful receipt of the message. It was successful and in 1980 the DIX Consortium (Digital Equipment Corp., Intel, and Xerox) was formed, issuing a specification, *Ethernet Blue Book 1*, followed by *Ethernet Blue Book 2*. This was offered to the Institute of Electrical and Electronics Engineers (IEEE, www.ieee.org), who in 1983 issued the *Carrier Sense, Multiple Access/Collision Detect* (CSMA/CD) specification–their stamp of approval on the technology.

Ethernet has since evolved under IEEE to encompass a variety of standards for copper, fiber, and wireless transmission at multiple data rates.

Ethernet is an excellent transmission medium for data, but by itself falls short of offering a complete solution. A network protocol is also needed to make it truly useful and what has evolved alongside of Ethernet is TCP/IP.

The big push toward TCP/IP came in the mid-1980s when 20 of the largest U.S. government departments, including the U.S. Department of Defense, decreed that all mainframes (read: expensive computers) to be purchased henceforth required a commercially listed and available implementation of UNIX to be offered. The department didn't necessarily need to use UNIX for the project at hand, but after "the project" was completed, the government wanted the ready option to convert this expensive computer into a general-purpose computer.

This soon meant that all serious computer systems in the world had relatively interoperable Ethernet and TCP/IP implementations. So IBM had Systems Network Architecture (SNA), TCP/IP, and Ethernet on *all* of its computers. Digital (DEC) had DECnet, TCP/IP, and Ethernet on all of its computers. Add a few more examples (Cray, Sun, CDC, Unisys, etc.) and you soon see that the only true standard available on all computers was a TCP/IP plus Ethernet combination.

Both from a historical view as well as in today's industrial world, the TCP/IP plus Ethernet marriage is a key combination. Neither would have survived or prospered without the other.

2.0–A Brief Tutorial on Digital Communication

Digital communication is the transmission of data between two or more intelligent devices in a mutually agreed upon electronic format (e.g., binary, octal, EBCDIC, and ASCII). The following components are necessary to accomplish this:

- Data source
- Transmitter
- Communications channel
- Receiver
- Data destination

The fundamentals of communication are the same, regardless of the technology. Confusion about any aspect can usually be helped with direct analogies to more familiar modes of communication such as multiple people engaged in a conversation around the dinner table, telephones, CB radios, or Morse code.

Communication standards define agreement on key details:

1. *Physical Connections*: How the signal gets from one point to another
   - The actual form of the physical connections
   - Signal amplitudes, grounding, physical media (coaxial cable, fiber, twisted pair, etc.)
   - Transmitting, receiving, and isolation circuitry
   - Safe handling of fault conditions such as miswiring or shorts to ground

2. *Coding*: How the message is represented by the 1's and 0's
   - Format of data units (e.g., ASCII)
   - Binary encoding of data (Manchester, RZ, NRZ, etc.)

3. *Protocol*: How messages are formatted and delivered
   - Error detection
   - Data flow control
   - Message prioritization
   - Time-outs: what happens when a response is not received
   - Synchronization: coordinating the timing of message events

2.1 Digital Communication Terminology

Signal Transmission

In the literal sense, all communication signals in transmission are analog. Whether it's a digital pulse train on a wire or laser light on a fiber-optic line, the physical nature of the media impose *attenuation* and *bandwidth* limits on the signal.

Attenuation

Loss of signal amplitude for any reason is called *attenuation*. When another person shouts to you from down the street, the farther away they are, the more attenuated their voice becomes—meaning loudness and clarity are lost. Attenuation is, among other things, a function of frequency and distance.

In combination with noise, attenuation dictates your ability to move data over long distances. The various digital communications standards are impacted by attenuation to different degrees. It is part of the reason that while RS-232 is limited to 15 m, RS-485 can go up to 1000 m. Whether you're talking about sounds in the air or signals on a wire, attenuation is normally expressed in decibels.

Bandwidth

The *bandwidth* of a medium is its ability to move useful data over time—for example, 400,000 words per second or 20 messages per second. Your transmission speed is dependent upon your bandwidth, while noise, data retries, and other "overhead" eat into your bandwidth.

Transmission speed is limited by the ability of the medium to rapidly change states between 1 and 0. When a transmitter changes from 0 to 1, there is some delay in the remote receiver noticing this change. Noise filtering causes the remote receiver to ignore the leading edge of the change, and capacitance and other electrical properties resist the change within the media.

The shouting example above also points the inverse relationship between attenuation and bandwidth—people naturally slow down their words as they shout to help compensate for the loss of sound.

Some transmission mechanisms use clever coding to greatly reduce the amount of bandwidth (high frequency content) required by high bit rates.

Noise

Noise is any unwanted signal that interferes with data transmission. Noise on a network can be created by external sources such as power lines, radios, welders, switchgear, cellular telephones, etc. and is induced by coupling of magnetic and/or electric fields. The most typical measurement of noise is the signal-to-noise ratio, expressed in decibels. The oft-cited advantage of digital communication is that if noise levels are kept below a certain threshold, it does not affect communication. In reality, noise is usually sporadic, sometimes affecting messages and sometimes not. Tips for reducing noise in Ethernet networks are given in Chapter 7, "Installation, Troubleshooting, and Maintenance Tips."

Encoding Mechanisms

When you think of 1's and 0's on a wire, it's intuitive to assume that the data appears on the wire exactly as it does in the packet. Actually this is seldom the case. During a very long string of continuous 0's or 1's (which is certain to happen from time to time), the receiver may think the connection has been lost and can lose synchronization with the transmitter. There are a variety of specific mechanisms for preventing problems like this, with tradeoffs between noise immunity, bandwidth, and complexity. The following are the most common formats:

- Manchester (used in 10-Mb Ethernet): The state of a bit is represented by a *transition* between V+ and V- in the middle of the bit. 1's are represented by a downward swing from V+ to V-; 0's are represented by an upward swing from V-to V+. There is ALWAYS a transition, regardless of the actual bit sequence. Advantage: The receiver and transmitter clocks are always synchronized. Disadvantage: This scheme uses twice as many transitions as bits.

- RZ (Return to Zero): The signal state is determined by the voltage during the first half of each bit, and the signal returns to a resting state ("zero") during the second half of each bit.

- NRZ (Non Return to Zero): This is simply a direct, intuitive, "1 = high, 0 = low" designation with no further coding.

- MLT-3: A three-level algorithm (i.e., high, zero, and low voltages) that changes levels only when a 1 occurs. Not self-clocking.

- Differential Manchester: Bit value is determined by the presence or absence of a transition at the beginning of a bit interval; clocking is provided via a mid-interval transition.

- 4B/5B (4 bit/5 bit): Every four bits is represented as a 5-bit code that never has more than three 0's in a row. This prevents long sequences of 0's or 1's with only a 25% penalty in bandwidth, in contrast to the 100% penalty of Manchester.

Signaling Types

The rubber meets the road in an Industrial Ethernet Network when the data from a device is transferred to the communication media. The process of transferring data to the wire is called *signaling* and there are two basic types: baseband and broadband.

Baseband signaling is digital. The 1's and 0's of a message are transmitted over the media as a sequence of voltage pulses. If you remember the old time westerns, the clerk in the telegraph office would tap out a telegram using digital signaling. The major limitation to digital signaling is that only a single message can be transmitted at a time. If there are two clerks in that telegraph office, the second clerk must wait for the first clerk to finish before sending the next message. A second limitation is that the digital voltages are easily attenuated as the distance increases. Communication over very large distances requires repeaters and is almost impractical due to the number of repeaters required.

Broadband signaling doesn't have this restriction. Broadband signaling uses analog carrier to transport data. Multiple carriers each containing data is analogous to a cable-TV system. The cable carries many television programs all on the same wire; all at the same time. You select a different carrier and a different data stream (a television program) by switching from Channel 9 to Channel 10.

BASEBAND	BROADBAND
Digital Signaling	Analog Signaling
Limited Distance	Long Distances
Bus-Oriented Applications such as RS-232 and Controller Area Networking (CAN)	Used for both Bus and Tree Topologies such as Token Ring and Ethernet
Bi-directional	Uni-directional
Single Message Oriented	Multiple carrier signals with multiple independent data streams
Often uses Manchester Encoding	No Encoding of digital signals

Error Detection

The simplest mode of error detection is "echoing back" the message just sent. However this consumes double bandwidth. Plus if there's an error, it's impossible to tell whether it was the original or the copy that was corrupted.

Checksum

The *checksum* calculation is effective for small amounts of data. An algorithm converts the data to bits that are appended to the data and transmitted. The receiver does the same calculation on the same data, if its own result does not match the original checksum, a retransmit request is submitted. For a single byte of data, a 1-bit checksum (parity bit) is sufficient.

Cyclic Redundancy Check

Long messages require a more sophisticated, more accurate detection method. *Cyclic Redundancy Check* (CRC) views the entire message block as a binary number, which it divides by a special polynomial. The result is a remainder, appended to the message just like a checksum. CRC calculation is performed in real time by logic gates at the hardware level.

Not only are the above mechanisms employed in hardware, they are also employed in higher-level protocols. TCP/IP employs its own error-detection mechanisms to further guarantee successful message delivery.

2.2 What's the Difference Between a Protocol and a Network?

The distinction between the physical network itself and the protocol that runs on that network is sometimes blurred. It's important to clarify: The network itself consists of the physical components and message-transmission hardware. Protocols are binary "languages" that run on the networks.

Strictly speaking, the terms *Ethernet, RS-232, RS-422, and RS-485,* for example, refer to the network physical wiring and message-transmission components only (see layers 1 and 2 of the ISO/OSI model in Chapter 3).

Many different protocols are used on Ethernet. TCP/IP, FTP, HTTP, NetBEUI, AppleTalk, and Modbus are protocols only; they can run on many different physical networks.

Transmission/Reception of Messages

Simplex. Simplex is one-way communication via a single channel. A radio or TV tower is a simplex transmitter; a radio or TV is a simplex receiver.

Duplex. Duplex is two-way communication.

Half Duplex. Half-duplex communication is when both stations (e.g., Walkie-Talkie or CB radio) can transmit and receive but they cannot do it simultaneously.

In half-duplex communication, only one party can have control of the channel at any one time. This necessitates an arbitration mechanism to determine who has control of the channel. This is called *contention*.

Full Duplex. Full duplex is two-way communication with two communications channels so that both stations can receive and transmit simultaneously. A telephone is full duplex because it allows both parties to talk and listen at the same time.

Figure 1 – Simplex Communication.

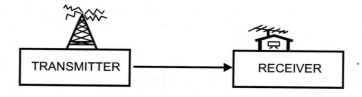

Figure 2 – Half Duplex Communication.

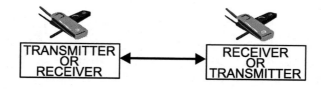

Figure 3 – Telephone conversation is full duplex (at least to the extent that a person can talk and listen at the same time).

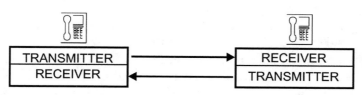

In Ethernet, half-duplex communication requires the use of CSMA/CD arbitration; full duplex eliminates collisions altogether but requires separate transmit and receive paths between each device. Ethernet always has separate Tx/Rx paths. Full Duplex in Ethernet requires not only separate paths, but only 2 nodes in a collision domain.

2.3 Basic Topologies

Topology is a very important choice in system design. It dictates what kind of physical arrangement of devices is possible. Figures 5 through 9 show what topologies are supported by each flavor of Ethernet.

Figure 4 – As bit rate increases, the physical length of each bit decreases.

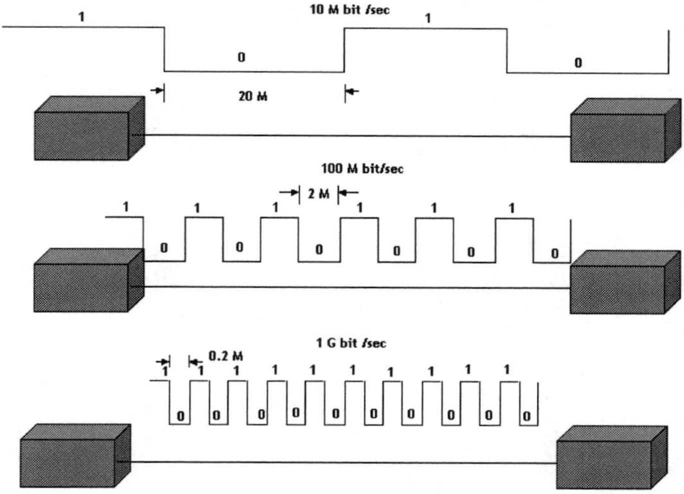

10 M bit /sec

1   1   0   0   20 M

100 M bit/sec

2 M   1   1   1   1   1   1   0   0   0   0   0   0

1 G bit /sec

0.2 M   1   1   1   1   1   1   1   1   1   1   0   0   0   0   0   0   0   0   0   0

Network speed	10 Mbps	100 Mbps	1 Gbps
Distance signal travels in the duration of 1 bit time	20 m	2 m	0.2 m

A network is an electrical transmission line. At high speeds, each bit is short compared with the network length. If you could physically see the packets traveling across the wire, each bit would have a length, similar to the wavelength of sound or light. Data propagates on wire at about 2/3 the speed of light.

When a wave reaches the end of a medium, it is reflected, transmitted, and/or absorbed. The shorter the bits in relation to the network, the more likely that reflections will cause errors.

For high-speed networks, the simplest way to minimize reflections is to have only one node at each end of a wire, with proper impedance termination (wave absorption) at each end. If each node has terminating resistors matching the cable impedance, reflections are minimized.

Hub/Spoke or Star Topology

A *hub/spoke* or *star topology*, where every segment has dedicated transmitters and receivers, offers high performance because reflections and impedance mismatches are minimal. This is the topology used by all of the Ethernet formats except 10BASE2 and 10BASE5.

Figure 5 — Star topology.

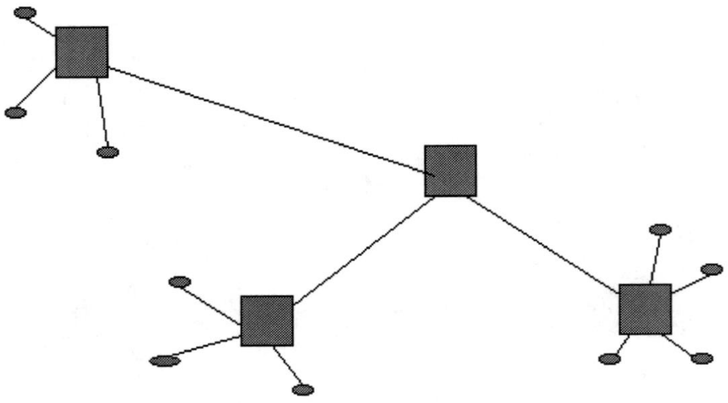

A BRIEF TUTORIAL ON DIGITAL COMMUNICATION

Ring Topology

Ring topology could be looked at as a variation on hub/spoke. It is similar in the sense that each segment has dedicated transmitters and receivers. However, the data itself is passed around in a circle, and it is stored and forwarded by each node – an important distinction.

Figure 6 – Ring topology.

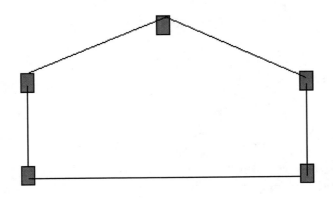

Mesh Topology

Mesh topology is point-to-point like star and ring, but has a minimum of two paths to and from each network node. This provides redundancy but introduces significant cost and installation effort.

Figure 7 – Mesh topology.

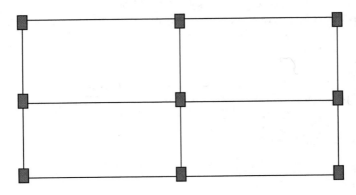

Trunk/Drop (Bus) Topology

Trunk/drop topology, also known as *Bus* or *Multidrop topology*, puts multiple nodes along the distance of the cable, with spurs or "Tees" inserted wherever a node is needed. Each spur introduces some reflections, and there are rules governing the maximum length of any spur and the total length of all spurs. 10BASE5 is a trunk/drop implementation of Ethernet.

Figure 8 — Bus topology.

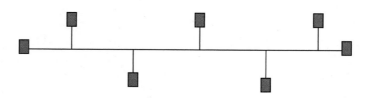

Daisy Chain Topology

A variation on trunk/drop is the *daisy chain*, where spur length is reduced to zero. High bandwidth signals have fewer problems in a daisy chain than trunk/drop because of fewer reflections. RS-485 is an example of a daisy chain; Controller Area Networks (CANs) like DeviceNet use trunk/drop. 10BASE2 is a daisy chain implementation of Ethernet; the drop length is effectively zero.

Star topologies have a nice advantage over trunk/drop: Errors are easier to assign to a single segment or device. The disadvantage is that some physical layouts (e.g., long conveyor system with evenly spaced nodes) are difficult to implement on a star; trunk/drop or daisy chain are better for that.

Figure 9 — Daisy chain topology.

A BRIEF TUTORIAL ON DIGITAL COMMUNICATION

2.4 Arbitration Mechanisms

There are three basic methods of arbitrating between competing message sources:

Contention

Contention is similar to a group of people having a conversation where all are listening, one can speak at any given time, and when there is silence another can speak up. Two or more may interrupt the silence and then all but one must back off and wait their turn.

Token

Token messaging is when each device receives some sort of token or "turn to speak" and can transmit only while it is in possession of that token. The token is then passed to someone else who now can transmit. Since Ethernet is not token-based, not much space will be given to this topic. There are many possible rules for passing the token, but often it is passed in a cyclic fashion from one device to the next.

Polling

Polling is when one device is "in charge" and asks each device to surrender its data in turn. Polling systems are often deterministic but do not allow urgent messages to be prioritized over other messages.

2.5 LAN vs. WAN vs. VPN

Local area networks (LANs) transmit data at high speed over a limited area. A single Ethernet system on 10BASE-T or 100BASE-T is a very typical LAN architecture. Such a system is limited in geography by the maximum number of hubs/switches (see Chapter 3), and propagation delays are in the 1-ms range and below.

Wide area networks (WANs) link LANs together over large distances. WANs usually use publicly available communication links from telecommunication providers. These links might consist of combinations of fiber, telephone, radio, and satellite links. Within a WAN, gateways often buffer packets until messages are complete, then forward them to the receiving computer. This causes propagation delays, and WANs are often unsuitable for real-time applications.

Virtual Private Networks (VPNs) link LANs via the Internet. Since data is then visible to others, data encryption is used to keep messages private. VPNs are extremely popular in companies with facilities in multiple locations and in companies that have remote or traveling employees. VPNs can extend through dial-up modem connections with the appropriate software installed on the dialing PC.

3.0—Ethernet Hardware Basics

3.1 Ethernet Terminology

The many formats of Ethernet cabling are described with rather unfriendly shorthand terminology. IEEE's Ethernet naming convention works like this:

- The first number (10, 100, 1000) indicates the transmission speed in megabits per second.

- The second term indicates transmission type: BASE = baseband; BROAD = broadband.

- The last number indicates segment length. A 5 means a 500-m segment length from original Thicknet.

   Tip 1 – **You might assume that the 2 in 10BASE2 indicates a 200-m segment length, but don't be too literal.** Actually 10BASE2 supports 185 m, or 300 m running point-to-point without repeaters.

- In the newer standards, IEEE used letters rather than numbers. The T in 10BASE-T means Unshielded Twisted-Pair cables. The T4 in 100BASE-T4 indicates four pairs of Unshielded Twisted-Pair cables.

10BASE5: Thick Ethernet (Thicknet)

10BASE5 is the original IEEE 802.3 Ethernet. 10BASE5 uses thick yellow coaxial cable with a 10-mm diameter. The cable is terminated with a 50-ohm 1-W resistor. One hundred stations maximum per segment are allowed.

10BASE5 uses trunk/drop topology. Stations are connected with a single coaxial cable. The maximum length of one segment is 500 m, limited by the quality of the cable itself.

A network interface card (NIC) is attached with a 15-pin D-shell connector to a short Attachment Unit Interface (AUI) cable, which in turn connects to a Media Attachment Unit (MAU) and links to the coaxial cable by means of a "vampire connector," which pierces the cable. The MAU contains the actual transceiver that connects to the coaxial cable.

For proper CSMA/CD operation, the network diameter for 10BASE5 is limited to 2500 m, consisting of five 500-m segments with four repeaters.

10BASE2: Thin Ethernet (THINNET)

10BASE2 resembles 10BASE5. It was introduced to reduce the cost and complexity of installation. It uses RG-58 50-ohm coaxial cable that is cheaper and thinner than that used for 10BASE5, hence the name Cheapernet or Thinnet, which is short for "Thin Ethernet."

10BASE2 integrates the MAU and the transceiver/AUI cable onto the NIC itself, with a Bayonet Nut Connector (BNC) replacing the AUI or D-15 connector on the NIC.

Lower cable quality means reduced distance. The maximum length of a 10BASE2 segment is 185 m. 10BASE2 supports 30 nodes per segment and keeps the four repeater/five segment rule. So the maximum network diameter of 5 segments × 185 m = 925 m.

Thinnet became very popular, at the time replacing Thick Ethernet as an office cabling solution.

10BASE-T: Twisted-Pair Ethernet

In 1990, IEEE approved 802.3i 10BASE-T, a completely new physical layer. It is very different from coax. 10BASE-T uses two pairs of Unshielded Twisted-Pair (UTP) telephone-type cable: one pair of wires to transmit data, and a second pair to receive data. It uses eight conductor RJ-45 connectors.

The topology is star instead of trunk/drop, with only two nodes per segment allowed: Station to repeater, repeater to repeater, or station to station with a *crossover cable*, which is needed to cross the transmit and receive lines.

The maximum length of a segment is 100 m, which follows the EIA/TIA 568 B wiring standard. Repeater-repeater links are also limited to a maximum of 100 m.

10BASE-T uses the four repeater/five segment rule from 10BASE5 and 10BASE2. So a 10BASE-T LAN can have a maximum diameter of 500 m.

Like 10BASE2 and 10BASE5, 10BASE-T uses Manchester encoding. IT uses +V and –V voltages with differential drivers. The signal frequency is 20 MHz, and Category 3 or better UTP cable is required.

10BASE-T has a link integrity feature, which makes installing and troubleshooting much easier. Devices on each end of the wire transmit a "heartbeat" pulse. Both the hub and the NIC look for this signal when connected. The presence of a heartbeat means a reliable connection is in place.

Most 10BASE-T devices have a light-emitting diode (LED) that indicates whether the link is good.

 ***Tip 2* – You should start troubleshooting wiring problems by looking at the state of the link LED at both ends of the wire.**

Most 10BASE-T equipment combines the functions of the MAU in the NIC or the hub itself.

In terms of bandwidth, coaxial cable is superior to UTP cable. However, UTP cabling and star topology are a real advantage because (1) in a bus topology, a problem at one node can take down the whole network, whereas a star topology makes it easier to isolate problems; (2) with the low cost of hubs and switches, star topology is still cost-effective, and (3) the existing cabling used in telecommunications equipment, especially CAT3 cable, could be used.

The star-shaped, planned, and structured wiring topology of telecommunications with 10BASE-T is very different and far superior to the single-point-of-failure method of 10BASE5 and 10BASE2.

10BASE-F: Fiber-Optic Ethernet

10BASE-F actually refers to three different sets of fiber-optic specifications:

- 10BASE-FL (FL means "Fiber Link") replaces the older Fiber Optic Inter-Repeater Link (FOIRL) spec and is backward compatible with existing FOIRL devices. It is the most popular 10-Mbps fiber standard and connects DTEs, repeaters, and switches. Equipment is available from many vendors.

- 10BASE-FP and 10BASE-FB are dead. P stands for Passive and B stands for backbone.

10BASE-F comes from the FOIRL specification of 1987, which linked repeaters using an extended distance fiber-optic link.

10BASE-F has twin strands of single-mode or multimode glass fiber, using one strand to transmit and the other to receive. Multimode fiber (MMF) of 62.5-/125-micron diameter is most often used with 10BASE-F to carry infrared light from LEDs. The specified connectors are IEC BFOC/2.5 miniature bayonet connectors, best known as ST connectors. SC and ST connectors are extremely popular in 10BASE-F.

Segment length for 10BASE-F ranges from 400 to 2000 m with a maximum of 5 segments on one collision domain.

10BROAD36 uses radio frequency transmission to carry data. This permits multiple channels to operate simultaneously on the same cable. 10BROAD36 is essentially dead, and no 100-Mbps version exists.

Fast Ethernet

100BASE-T. 100BASE-T is basically 10BASE-T with the original Ethernet Media Access Controller (MAC), at 10 times the speed. The 100BASE-T allows several physical layer implementations. Three different 100BASE-T physical layers are part of IEEE 802.3u: two for UTP and one for multimode fiber. Just like 10BASE-T and 10BASE-F, 100BASE-T requires a star topology with a central hub or switch.

IEEE 802.3u contains three new physical layers for 100-Mbps Ethernet:

- **100BASE-TX:** Two pairs of Category 5 UTP or Type 1 STP cabling; most popular for *horizontal* connections. Uses two strands of 62.5-/125-micron Fiber Distributed Data Interface (FDDI) cabling.

- **100BASE-FX:** Two strands of multimode fiber; most popular for *vertical* or backbone connections.

- **100BASE-T4:** Four pairs of Category 3 or better cabling; not common. 100BASE-T4 was part of IEEE 802.3u and was intended to capitalize on the huge installed base of Category 3 voice-grade wiring. It was a flop because T4 products only started shipping a year after the standard was approved; TX products got a head start.

See Table 3-1 for distance capabilities.

ETHERNET HARDWARE BASICS

Table 3-1 – Ethernet Physical Layer Characteristics

Format	Data rate	Max segment length *	Max nodes per segment	Topology	Media	Connectors	Encoding	Notes
10BASE-T	10 Mbps half duplex 20 Mbps full duplex	100 m Max network length = 100 m node to hub	2	Star	Category 3, 4, or 5 UTP cable with two pairs of voice-grade/ telephone twisted pair, 100 ohms	8-pin RJ-45 style modular jack; industrial variants include M18, M12, and DB9	Manchester	Most popular 10-m format
10BASE2 "Thinnet" or "Cheapernet"	10 Mbps half duplex only	185 m Max network length = 925 m = 5 × 185 m	30	Bus with drops. Minimum spacing between nodes = 0.5 m, max drop length = 4 cm	5-mm "thin" coax, e.g., RG58A/U or RG58C/U, Belden 9907 (PVC), and 89907 (plenum); 50 ohms	BNC "T" coax connectors, barrel connectors, and terminators	Manchester	5-cm min bend radius; may not be used as link between 10BASE5 systems
10BASE5 "Thicknet"	10 Mbps half duplex only	500-m (50-m max AUI length) Max network length = 2800 m = 5 × 500 m segments + 4 repeater cables + 2 AUI cables	100	Bus with drops	10-mm ("thick") coax, e.g., Belden 9880 (PVC) and 89880 (plenum); bend radius min 25 cm; 50 ohm media and termination	N-type coaxial connectors, barrel-style insulation displacement connectors and terminators	Manchester	MAU links trunk to NIC via AUI cable; taps must be spaced at 2.5-m intervals; ground at one end of cable**

Table 3-1 – Ethernet Physical Layer Characteristics (continued)

Format	Data rate	Max segment length *	Max nodes per segment	Topology	Media	Connectors	Encoding	Notes
10BROAD36	10 Mbps half duplex only	1800-m single segment; 3600 m total for multiple segments			75-ohm CATV broadband cable		Modulated RF	Dead
10BASE-FL	10 Mbps half duplex; 20 Mbps full duplex	2000 m	2	Star	2 MMF cables, RX and TX, typically 62.5/125 fiber, 850-nm wavelength	BFOC/2.5, also called "ST"	Manchester	Uncommon
100BASE-TX	100 Mbps half duplex; 200 Mbps full duplex	100 m	2	Star	2 pairs of Category 5 UTP cabling; 100-ohm impedance (optionally supports 150-ohm STP)	RJ-45 style modular jack (8 pins) for UTP cabling (optionally supports 9-pin D-shell connector for STP cabling)	4B/5B	Most popular 100-m format IEEE 802.3u
100BASE-FX	100 Mbps half duplex; 200 Mbps full duplex	Half duplex: 412 m; full duplex: 2000 m	2	Star	2 MMF optical channels, one for TX, one for RX. Typ. 62.5/125 MMF, 1300-nm wavelength	Duplex SC, ST, or FDDI MIC connectors	4B/5B	IEEE 802.3u

Table 3-1 – Ethernet Physical Layer Characteristics (continued)

Format	Data rate	Max segment length *	Max nodes per segment	Topology	Media	Connectors	Encoding	Notes
100BASE-T4	100 Mbps half duplex only	100 m	2	Star	Category 3, 4, or 5 UTP (uses 4 pairs or wires); 100 ohm	RJ-45 style modular jack (8 pins)	8B/6T	Uncommon IEEE 802.3u Useful where existing CAT3 telecom cables are available
1000BASE-LX	1000 Mbps half duplex; 2000 Mbps full duplex	Half-duplex MMF & SMF: 316 m; full-duplex MMF: 550 m; full-duplex SMF: 5000-m; 10-micron SMF: 3000-m max segment length	2	Star	2 62.5/125 or 50/125 multi-mode optical fibers (MMF), or 2 10-micron single-mode optical fibers (SMF), 1270- to 1355-nm light wavelength	duplex SC connector	8B/10B	803.z Multimode: longer-building backbones; Single mode: campus-wide backbones
1000BASE-SX	1000 Mbps half duplex; 2000 Mbps full duplex	Half-duplex 62.5/125: 275 m; half-duplex 50/125: 316 m; full-duplex 62.5/125: 275 m; full-duplex 50/125: 550 m	2	Star	2 62.5/125 or 50/125 MMF, 770 to 860 nm	duplex SC connector	8B/10B	
1000BASE-SX	1000 Mbps half duplex; 2000 Mbps full duplex	Half duplex: 25 m; full duplex: 25 m 62.5/125 MMF full duplex: 260 m	2	Star	Specialty shielded balanced copper jumper cable ("twinax" or "short haul copper")	9-pin shielded D-subminiature connector, or 8-pin ANSI Fibre Channel Type 2 (HSSC) connector	8B/10B	802.3z Intended for short backbones

Table 3-1 – Ethernet Physical Layer Characteristics (continued)

Format	Data rate	Max segment length *	Max nodes per segment	Topology	Media	Connectors	Encoding	Notes
1000BASE-T	1000 Mbps half duplex; 2000 Mbps full duplex	100 m (328 ft)	2	Star	4 pairs of CAT5 or better cabling, 100 ohms	8-pin RJ-45 connector	PAM5	802.3ab Replace existing 10/100BASE-T runs in floors of buildings
1000BASE-CX		25 m	2	Star	STP copper Twinax, 150 ohms	DB9 or HSSC	8B/10B	802.3z Short jumper connection in computer rooms or switching closets Common ground required on devices at both ends of the cable

AUI = Attachment Unit Interface; MAU= Medium Attachment Unit; MMF = Multimode fiber; NIC = Network interface card; SMF = Single-mode fiber; STP = Shielded Twisted Pair; UTP = Unshielded Twisted Pair; 62.5/125 means 62.5-micron fiber core with 125-micron outer cladding.
* For best results, keep segment length at least 20% shorter than recommended maximum.
** Ground should be made at one and only one point in a single link.
Maximum transmission path rules: 5 segments, 4 repeaters, 3 coax segments OR 5 segments, 4 repeaters, 3 link segments, 2 coax segments.

ETHERNET HARDWARE BASICS

Gigabit Ethernet

1000-Mb Ethernet is just Fast Ethernet on steroids. 100BASE-T was wildly successful and it was only a matter of time before the data rate would be increased again. There are differences in the physical layers, network design, and minimum frame size. IEEE 802.3z, approved in 1998, includes the Gigabit Ethernet MAC, and three physical layers. Gigabit uses 8B/10B encoding. Gigabit encompasses three physical standards:

- 1000BASE-SX Fiber

- 1000BASE-LX Fiber

- 1000BASE-CX Copper

- 1000BASE-T

Some engineers wanted Gigabit Ethernet to be full duplex only, but CSMA/CD was kept. One of the reasons for keeping it was that it reduced the amount of redesign in migrating to 1000-Mb chips.

To make CSMA/CD work at 1 GHz, the slot time was increased to 512 bytes, as opposed to 64 bytes for 10- and 100-Mbps Ethernet. This is the allowable time in which the transmitter "holds the floor" and the complete frame is transmitted. If the transmitted frame is smaller than 512 bytes, a carrier extension is added to the end of the frame. The carrier extension resembles the PAD (see later in this chapter) that is added to the end of the data field inside the frame. Carrier extension, however, adds after the CRC and does not actually form part of the frame.

If most of the messages on a gigabit network are short, overhead makes the network extremely inefficient. So Gigabit Ethernet includes a feature called *burst mode*. A station can continuously transmit multiple smaller frames, up to a maximum of 8192 bytes. This is done such that the transmitting node has continuous control of the media during the burst.

1000BASE-SX: Horizontal Fiber. 1000BASE-SX is for low-cost, short-backbone, or horizontal connections (*S* stands for "short"). It has the same physical layer as LX and uses inexpensive diodes and multimode fiber. Distance ranges from 220 to 550 m, depending on the type of fiber.

1000BASE-LX: Vertical or Campus Backbones. 1000BASE-LX is for longer-backbone and vertical connections (*L* stands for "long"). LX can use single-mode or multimode fiber. It requires more expensive optics. Segment length is 5000 m with single-mode fiber. For full-duplex multimode, the distance is 550 m. IEEE specifies the SC style connector for both SX and LX.

1000BASE-CX: Copper-Twinax Cabling. 1000BASE-CX (*C* stands for "copper" or "cross-connect") links hubs, switches, and routers in closets. Copper is preferred because it is faster to wire a connection with copper than with fiber. 150-W twinax cable is specified. Maximum length is 25 m for half or full duplex. Two connectors are used with 1000BASE-CX: The High-Speed Serial Data Connector (HSSDC) and the 9-pin D-sub-miniature connector, used for token ring and the 100BASE-TX STP.

 Caution: You should be aware that Fast Ethernet and Gigabit Ethernet systems on copper media are extra susceptible to electrical noise for two reasons: (1) voltage levels are lower and thus more easily corrupted by noise, and (2) bit times are extremely short; a noise spike doesn't have to last very long to corrupt an entire frame.

 Tip 3 – Take extra care to select high-quality cables and avoid routing them through electrically noisy areas if possible.

3.2 Ethernet Hardware LEDs

Most Ethernet NIC cards, hubs switches, and other hardware have two LEDs:

1. The Link LED indicates that a reliable physical connection is established between the device and another device. This LED is the very first thing you should check when something doesn't appear to be working correctly.

2. The TX LED turns on when the device is transmitting data. Some devices also have an RX LED that indicates data is being received.

Tip 4 – These LEDs are the first thing you check when there appears to be a problem.

3.3 Physical/Embedded Components: MAC, PHY, and Magnetics

At the lowest level, an Ethernet interface typically is made of four components. The first two are usually combined:

- *Media Access Controller* (MAC): For example, the popular AMD 79C960 chip and its derivatives. The Controller assembles and disassembles Ethernet frames and provides an interface to external data.

- Internal to most MACs (separate in some cases) is an intermediate interface. It allows independence from the different types of transmission media (copper, fiber). For 10-Mbps Ethernet, this is called an *Attachment Unit Interface* (AUI). In 10BASE5 systems, this is an external piece of hardware. In Fast Ethernet this is called a *Media-Independent Interface* (MII). The MII links the MAC and PHY chips. It allows different types of PHYs to be controlled by one MAC. 100-Mbps Ethernet calls this interface the *Media-Independent Interface* (MII), whereas Gigabit Ethernet calls it the *Gigabit Media-Independent Interface* (GMII).

- The *PHY* encodes data from the MAC, for example, Manchester or 4B/5B, and produces signal levels that can drive the magnetics and the cable.

- The *Magnetics* are an isolation transformer that protects the circuitry from voltage and current surges on the cable. They serve the same function that optical isolation serves in many other networks. Typical isolation is 1500 VDC but some industrial applications may require more.

Industrial-grade Ethernet hardware interfaces differ from office-grade gear in the following ways:

- **Common Mode Rejection Ratio:** at least 40 dB and as high as 60 dB

- **Higher surge protection ratings:** more than 2000 V instead of the standard 1500 V

- **More space between components** to prevent arcs

- **Transient protection circuitry** on transmit and receive sections

- **More copper on the circuit board** to reduce susceptibility to noise

Tip 5 – **When in doubt, take the lower-risk approach and select industrial-grade hardware.** The price difference is far less than "the real cost of not doing it right the first time."

3.4 Auto-Negotiation

Most NICs, hubs, and switches that support Fast Ethernet also support 10 Mb and automatically adjust their speed to match the node on the other end of the wire. This is tricky because there are seven possible Ethernet signals on an RJ-45 connector: 10BASE-T half or full duplex, 100BASE-TX half or full duplex, 100BASE-T2 half or full duplex, or 100BASE-T4.

Figure 10 – Components in a typical ethernet NIC card.

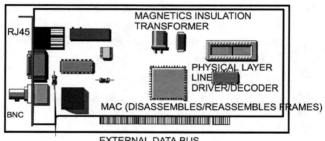

While the details of Auto-Negotiation are beyond the scope of this book, it's important to realize that this handy feature saves you *lots* of time. Without it, you would be forced to shuttle back and forth between nodes, making manual adjustments until each node was in agreement. Given the number of nodes and potential distances involved, this alone would seriously dampen the world's enthusiasm for Ethernet.

Auto-Negotiation logic is incorporated in nearly all equipment shipped after 1996. Auto-negotiation is an upgrade of 10BASE-T link integrity and is backward compatible with it.

Figure 11 – Collision domain analogy.

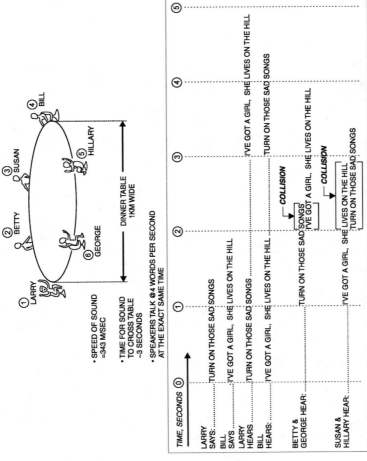

- SPEED OF SOUND
  =343 M/SEC

- TIME FOR SOUND
  TO CROSS TABLE
  ~3 SECONDS

- SPEAKERS TALK @ 4 WORDS PER SECOND
  AT THE EXACT SAME TIME

DINNER TABLE
1KM WIDE

LARRY AND BILL HEAR EACH OTHER WITH NO INTERRUPTION OR COLLISION, BUT THE MESSAGE IS GARBLED FOR THE OTHERS. TO ENSURE THAT NO COLLISION IS POSSIBLE, LARRY AND BILL MUST EACH SPEAK FOR A MINIMUM OF SIX SECONDS AND LISTEN FOR COLLISION. OTHERWISE THE POSSIBILITY EXISTS FOR SOMEONE TO RECEIVE A GARBLED MESSAGE.

3.5 Network Collisions and Arbitration: An Analogy

Imagine that you are having dinner with five other people, but the dinner table is 1 km wide instead of regular size. Assume for this illustration that you can easily hear each speaker despite the large distance.

Sound takes 3 s to travel 1 km. So if you and your friend across the 1-km table both start speaking at the same time, it will take 3 s before you know you are interrupting each other.

Successful contention rules would require the following conditions to be met:

- The rules that determine retransmit times must provide for at least 3 s of spacing between permitted transmissions.

- It will take each speaker at least 6 s to be certain that he or she is not being interrupted – if you start talking, your voice takes 3 secs to reach the other side. Assume the other guy starts talking 2.99 secs after you did. It will now take 3 secs for his voice to reach you. That means you have to listen out for the total "Round Trip Time" of 6 seconds.

- Therefore *each speaker must talk for more than 6 s every time he or she has something to say.* If two people simultaneously talked for only 2 s each, they would each hear the other's message clearly. The other speaker's message would arrive 1 s after he finished speaking and he could hear it, but others around the table would hear both messages mixed together.

- The larger the table is, the longer the messages must be if everyone has equal opportunity to talk.

- In this example, message length was described in seconds, not bits or bytes. There is a direct relationship between allowable network length and minimum message length. At this table, if people speak at a rate of 4 words per second, then the minimum message size is 24 words.

- Suppose the baud rate goes up—extremely talkative speakers appear, who speak 40 words per second instead of 4. Then the minimum message size is now 240 words! When you move from 10-Mb Ethernet to 100-Mb or 1-Gb, the minimum required message length grows. However to maintain compatibility, you cannot do this. So you have to reduce the size of the table. So from 10 Mbps to 100

Mbps the frame stayed at 64 Bytes min, so the collision domain shrank from 2500 m (51.2 uS) to 250 m (5.12 uS).

In Ethernet, a message must be long enough to reach the other end of the network before the transmitter stops transmitting. The minimum message length defines a maximum network length, which is called the *collision domain.*

3.6 How the CSMA/CD Protocol Works

Whenever you interrupted your sister at the dinner table, did your wise and all-knowing parents remind you that you have two ears but only one mouth? They were teaching you a basic principle of communication. Dinner conversation is a contention / collision detection mode of communication. When there is a lull in the conversation, someone who has something to say speaks and "has the floor."

In Ethernet terminology, he does not hear a Carrier signal from anyone else, and thus takes control of the network. Others who wish to speak must listen for a gap (**Carrier Sense**) and multiple people have the opportunity to take the next turn (**Multiple Access**). When there is silence and two people speak up at the same time, they both hear the interruption (**Collision Detect**) and, if they are polite, they will both stop speaking and wait their turn. One will choose to speak first and then she "has the floor."

This is exactly how Ethernet works in half-duplex mode. (CSMA/CD is *not* required in full-duplex mode.) Each node listens to the wire and if another node is transmitting, the other nodes remain silent until the channel is free (i.e., no carrier is sensed). When the bus is quiet, a node with data to transmit will send it.

It's quite possible that another node also has data to send, and it starts transmitting at the same time. A collision occurs, also detected by both nodes. They stop and choose a random number that indicates how long to wait to retransmit. The one with the lowest number re-tries first, and the message can now successfully be sent. The *back-off algorithm* for choosing this random number is designed to minimize collisions and re-transmissions, even when many, many nodes (up to 1024) are involved. A 10-Mb system generates backoff delay values ranging from 51 μs to 53 ms.

Table 3-2 – How an Ethernet Data Frame Is Constructed

Preamble	Start Frame Delimiter (SFD)	MACID of source	MACID of destination	TAG	Type / Length of data field	Message	PAD	CRC
Preamble (56 bits/7 bytes) 10101010---): used by the receiver to synchronize with the transmitter before actual data is passed	Start Frame Delimiter (SFD): (8 bits/1 byte. 10101011); indicates commencement of address fields	MACID of source (48 bits/6 bytes): 3 octets with NIC block license number – designates manufacturer + 3 octet device identifier	MACID of destination (48 bits/6 bytes; same format as source address)	*	Type / Length of data field 2 bytes	Message 0-1500 bytes	PAD 0–46 bytes Random "filler" data kicks in if message length is less than 46 octets to ensure minimum frame size of 64 bytes	CRC 4 bytes Cyclic Redundancy Check ("Frame Check Sequence," 32 bits/4 bytes)
			Three addressing modes: Broadcast: FFFFFFFFFFFF Multicast: First bit = 1 Point to Point: First bit = 0					

* TAG field in newer Ethernet frames adds 4 bytes for priority levels of quality of service. This supports advanced streaming and real-time services like VoIP (Voice Over Internet Protocol). Tag field includes 3 bits for priority and 11 bits for VLAN address. * This is an IEEE 802.3 frame. not an Ethernet V2 frame; there are subtle differences.

Notes: The shortest possible frame = 64 octets with a message length of 46 bytes, excluding SFD and preamble. Longest possible frame = 1518, excluding SFD and preamble. The receiver detects the existence of pad data on the basis of the value in the Length field. All higher-level protocols (i.e., TCP/IP and anything "riding on top") operate entirely within the message field. The 48-bit physical address is written in pairs of 12 hexadecimal digits, as 12 hex digits in pairs of 2.

ETHERNET HARDWARE BASICS

Tip 6 – Ways to Reduce Collisions:

• Minimize the number of stations on a single collision domain. Switches and routers divide a network into multiple domains.

• Avoid mixing real-time data traffic with sporadic "bulk data" traffic. If a network is handling regular cycles of I/O data, transferring a 10-MB file over the same network will compromise performance.

• Minimize the length of each cable.

• When possible, put high-traffic nodes close to each other.

• Use hubs and repeaters with buffers ("store and forward").

• Beware of "plug and play" devices, which may clog a network as they search for other devices on the network.

3.7 The Basic "Ethernet Design Rules"

The "5-4-3-2" rule states that the maximum transmission path is composed of 5 segments linked by 4 repeaters; the segments can be made of, at most, 3 coax segments with station nodes and 2 link [10BASE-FL] segments with no nodes between. Exceeding these rules means that some (though not all) nodes will be unable to communicate with some other nodes. You should check your design to ensure that no node is separated from any other node by more intermediate devices than the table below indicates.

Table 3-3 – Maximum Transmission Path Between Any Two Nodes

5 segments 4 repeaters 3 link segments 2 coax segments	OR	5 segments 4 repeaters 3 coax segments 2 link segments

Note: This table is a popular simplification of the actual 802.3 rules.

3.8 "Would Somebody Please Explain This 7-Layer Networking Model?" (Adapted from *Sensors Magazine,* July 2001, ©Advanstar)

Networks, and the information that travels on them, are most easily understood in layers. For many years the ISO/OSI model has been used as a way to represent the many layers of information in a network, particularly the low-level transport mechanisms. From top to bottom, these are the layers and how these layers relate to your product design (Table "Layer 1").

Please note that most networks do not actually use all of these layers, only some. For example, Ethernet and RS-232 are just physical layers—layer 1 only for RS-232 and layers 1 and 2 for Ethernet. TCP/IP is a protocol, not a network, and uses layers 3 and 4, regardless of whether layers 1 and 2 are a phone line, a wireless connection, or a 10BASE-T Ethernet cable.

Figure 12 – The 7-layer network concept.

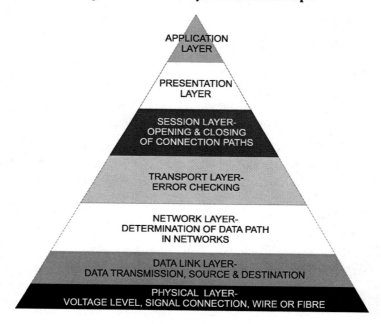

Layer 7: Application

The application layer defines the meaning of the data itself. If you send me a .PDF file via e-mail, the application that is used to open it is Adobe Acrobat. Many layers of protocols are involved, but the application is the final step in making the information usable.

In a sensor design, this is the software component that exchanges process data between the sensor elements (and their associated A/D converters, etc.) and the communications processor. It recognizes the meaning of analog and digital values, parameters, and strings.

ETHERNET HARDWARE BASICS

J1939 and CANOpen are application layers on top of CAN. Foundation Fieldbus HSE is an application layer on top of Ethernet and TCP/IP. Modbus is an application layer on top of RS-232/485.

Layer 6: Presentation

The presentation layer converts local data into a designated form for sending and for converting received data back to the local representation. It might convert a character set such as MacRoman to ASCII for transmission. Encryption can happen in this layer.

Layer 6 is usually handled by application software and is not usually used in industrial networks.

Layer 5: Session

The session layer creates and maintains communication channels (sessions). Security and logging can be handled here.

Layer 5 is handled by software and is not commonly used in industrial networks.

Layer 4: Transport

The transport layer controls transmission by ensuring end-to-end data integrity and by establishing the message structure protocol. It performs error checking.

Layer 4 is usually handled in software (e.g., TCP/IP).

Layer 3: Network

The network layer routes data from node to node in the network by opening and maintaining an appropriate path. It may also split large messages into smaller packets to be reassembled at the receiving end.

Layer 3 is done in software.

Layer 2: Data Link

The data link layer handles the physical transmission of data between nodes. A packet of data (data frame) has a checksum, source, and a destination. This layer establishes physical connection between the local machine and the destination, using the interface particular to the local machine.

Layer 2 is almost always done in hardware with Application Specific Integrated Circuits (ASICs). Low-speed networks can perform layer 2 functions in software.

Layer 1: Physical Layer

Layer 1 defines signal voltages and physical connections for sending bits across a physical media and includes opto-isolation, hubs, and repeaters. Physical media refers to the tangible physical material that transports a signal, whether copper wire, fiber, or wireless.

The data to be transferred starts out in the application layer and is passed down the seven layers to the physical layer, where it is sent to the receiving system. At that end, it is passed up through the layers to the remote application layer, where it is finally received by the user.

Just like a matrishka doll, when you encapsulate data at a particular layer, you must then wrap it in the lower layers from top to bottom; then when you unpack it, you must reverse the process.

Most protocols are related to the ISO/OSI model, but do not follow the exact specification. Instead, they combine different layers as necessary.

3.9 Connectors

You don't have to spend much time investigating industrial Ethernet to discover that RJ-45 "telephone connectors" aren't viewed with a great deal of respect. Nor should they. The design lacks even the most minimal environmental protection and can be easily damaged with a good yank on the cable. The surface area of the contacts is quite small and if the thin layer of gold over nickel is worn away by vibration, it becomes susceptible to corrosion and oxidation. Not a great choice for your robotic welder, especially if downtime costs $15,000 per minute.

Figure 13 – The ubiquitous RJ-45 connector for ethernet.

 *Tip 7 –* **Fortunately there are alternatives, three in particular from the industrial world.** They have been designed to keep out liquids (e.g., IP65 or IP67), maximize contact surface area, and improve the sturdiness of the design. All of them facilitate feeding Ethernet cables through panels, simply by choosing appropriate receptacles.

IP67 Sealed Connector System for Industrial Ethernet

Many applications in Medical, Aerospace, Food, and Pharmaceuticals require absolutely secure and trouble free communications. Other industrial applications are located in areas with extreme moisture, dirt, EMI and vibration. Standard office-grade RJ-45 connectors are not designed for these kinds of conditions. One of the ways to meet the requirements of these applications is to use sealed connector systems from vendors like Woodhead Industries, Phoenix Electronics and others.

There are several other connector options that are being promoted for industrial applications. These options are less rigorous than the RJ-Lnxx Woodhead connector described above. For example, PROFInet recommends an IP20 version of the RJ-45 connector from Harting. Other vendors recommend M12 connectors. The connector choice for a particular application must be matched to the environment. In some applications, an office-grade RJ-45 will work just fine. In more dirty or dusty environments, an IP20 or M12 type connector might be best while an IP67 type connector may be required for the most difficult environments.

Figure 14 – Woodhead Industries RJ-Lnxx System

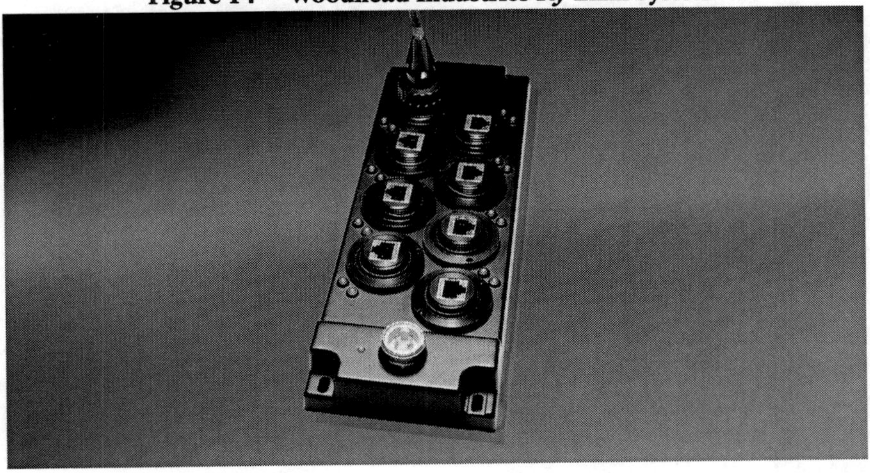

Figure 15 – Sealed RJ-45 Connector

3.10 Pinouts

Table 3-4 – RJ-45 from www.pin-outs.com

Pin No.	Function	Color
1	TX +	White/Orange
2	TX -	Orange
3	RX +	White/Green
4		Blue
5		White/Blue
6	RX -	Green
7		White/Brown
8		Brown

Figure 16 – RJ-45 standard connector pinout.

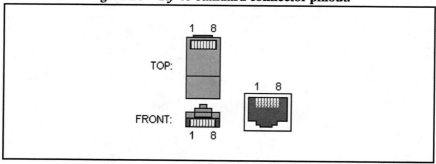

The color scheme for crossover cable is used when a hub or switch port does not flip the transmit and receive pairs. It also can be used to link one PC NIC card to another. This ensures that receivers talk to transmitters and transmitters talk to receivers.

Figure 17 — Color scheme for crossover cable.

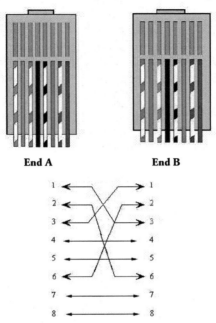

End A   End B

Table 3-5 — Category 5 Cabling Information (1000BASE-T Only). 1000BASE-T uses the standard registered-jack, RJ-45, connector.

connector RJ-45	pin#	Description	ANSI/TIA/EIA-568A
	1	Transmit Data1 + (TxD1+)	white/green
PIN 1	2	TxD1-	green/white
	3	Receive Data2 + (RxD2+)	white/orange
	4	RxD3+	blue/white
	5	RxD3-	white/blue
	6	RxD2-	orange/white
RJ-45M Male	7	TxD4+	white/brown
AT&T 258A - EIA/TIA 568B	8	TxD4-	brown/white

(From http://techsolutions.hp.com/dir_gigabit_external/training/techbrief6.html)

Ethernet DB-9 Connector

The trusty DB-9 has been employed for Ethernet systems, especially in Europe, and although few DB-9 designs are waterproof, it is certainly sturdier than an RJ-45.

Table 3-6 — Pinout for Ethernet Using Standard DB-9 Connectors

Pin No.	Function	Color
1	RX+	White/Green
2	-	-
3	-	-
4	-	-
5	TX+	White/Orange
6	RX-	Green
7		
8		
9	TX-	Orange

```
  1       5
 /○ ○ ○ ○ ○\
 \ ○ ○ ○ ○ /
  ‾‾‾‾‾‾‾‾‾
  6       9
```

female view

M12 "Micro" Connector for Industrial Ethernet

This is based on the ever-popular 12-mm "micro"/"euro" design, which is popular in automation. It has eight poles and uses four of them for the Ethernet signal. The other four can presumably be used for other purposes (e.g., power *a la* IEEE 802.3af), as wise practice and future standards dictate.

Figure 18 – Male and female pinouts for ethernet M12 connectors.

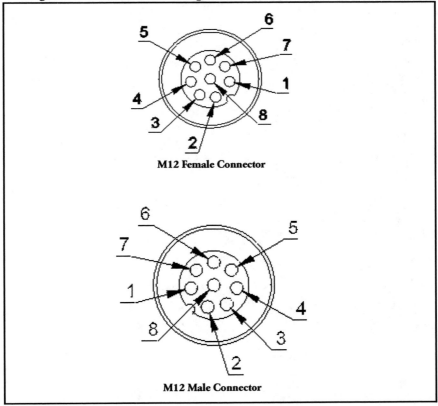

Pin No.	Function
1	-
2	-
3	-
4	TX-
5	RX+
6	TX+
7	-
8	RX-

4.0–Ethernet Protocol & Addressing

4.1 A Little Bit of History

Did you know that the networking cable connecting your computer to the Internet dates back to 1970? During that time period, a Harvard graduate student named Robert Metcalfe read a paper about something called Aloha Net. Aloha Net was a radio system used in the Hawaiian Islands to send small messages, also called *data packets*, between islands. A key feature of this network was that anyone could send messages at any time. If no acknowledgment was received, the message was not delivered and would be resent.

Dr. Metcalfe reasoned that with some mathematical enhancements to the system, the efficiency of the Aloha Net network, which then hovered at between 15% and 20%, could be vastly increased. Not only did the efficiency increase all the way up to 90%, but the packet communications network he designed became the worldwide standard we know today as Ethernet. Now known as IEEE Standard 802.3, it still retains the elegance and simplicity of the original Aloha Net.

Dr. Metcalfe later went on to found 3COM Corporation, one of the leading manufacturers of Ethernet adapter cards and a major communications company. His discovery spawned billions of dollars in global wealth. Today Ethernet continues to gain momentum as more than half of the world's computers are linked to an Ethernet network.

**Figure 19 – A replica of the diagram drawn by
Dr. Robert M. Metcalfe in 1976**

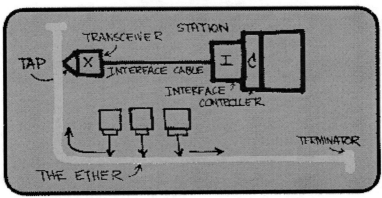

4.2 The Ethernet Packet and How Messages Flow on Ethernet

A somewhat simplified diagram of the contents of an Ethernet packet is shown in Figure 20. This packet contains two address fields, some data and a field that verifies correct reception of the packet.

Figure 20 – A Simplified View of an Ethernet Packet

Source Address	Destination Address	Packet Data	Error Checking Data

The first two fields are unique 48-bit addresses of the sending computer and the destination computer. These addresses are not the familiar "192.168.0.10" type addresses we often see but are addresses assigned by the manufacturers of the physical Ethernet cards. Every manufacturer producing Ethernet hardware is assigned a series of 48-bit addresses. Known as the MAC (Media Access Control) address, the manufacturers of Ethernet interface cards must ensure that they use only the addresses in their series and use it only once. That way no two computers in the world can be assigned the same address. This is much more rigorous than the addressing system used by the post office. In the post office there can be many destinations with identical addresses. There can be a "1118" on Main street, one on Elm, and many others in other cities. How these 48-bit addresses are translated to the familiar dotted-decimal addresses is a subject covered a bit later in this chapter.

You should not make the mistake of confusing the "destination" 48-bit address in an Ethernet packet with the final destination of your message. The source and destination in an Ethernet packet are simply the addresses of the sending and receiving computers in the chain of computers between your desk and your best friend 2,000 miles away. It is very much like taking a bus to work. In the middle of any two stops, the "source" address is the stop you just left, while your current "destination" is the next stop. Once you reach that stop, you can continue on to the next stop and continue with a new "source" and "destination", get off the bus and take a different one or begin work if you have arrived at your final destination.

In the same fashion, Ethernet packets simply flow from one computer to the next where one of three things can happen to the data field. First, it can be retransmitted to the next computer in the chain; akin to staying on the bus from our previous example. Second, it can be consumed by this computer if the final destination is this computer, or third, it can be discarded.

A message can be discarded for any number of reasons. It may be discarded if there is a checksum error on one of the several checksums in the message, if the message isn't delivered in a timely fashion or if the last field in the packet, the Error Checking field, indicates an error. Ethernet uses something called a CRC (Cyclic Redundancy Check) to verify accurate reception of a message. If the CRC is invalid, the message is rejected by your Ethernet interface and is never delivered to any of the TCP/IP protocol software on the receiving computer. In some cases, as we will see later, the sending protocol may expect a response and retransmit the message when the response isn't received or it may just ignore the lost message.

The CRC algorithm is a very accurate method to detect message errors. Statistically, it can only misinterpret messages a few times in every ten thousand messages.

The firmware in your Ethernet interface hardware that checks the destination address, controls access to the network and verifies the CRC is known as the MAC (Media Access Control). This layer is not part of the TCP/IP protocol stack in your computer. Instead, the MAC software is a firmware layer that is part of the hardware interface to Ethernet. In the past, it was typically a stand-alone computer chip that you could identify by looking at your Ethernet hardware. In many cases, the MAC chip is now included as part of the intelligence on board your Ethernet interface hardware.

One of the tasks for the MAC software is to monitor the network and insert a message on the network when no other messages are detected. This operation is very similar to driving and entering a highway from an onramp. Your car must fit into an open slot on the highway and if there isn't an open slot, you have to wait for one. Unlike the highway, however, there can be thousands if not millions of crashes. When a crash, or collision, happens the MAC software hears the "crunch" as the bits collide and reschedules the message for a random time in the future. Hopefully, there is an open slot at that time. On busy networks, just as on busy highways, waiting for an open slot can delay your trip considerably. The process of detecting message collisions, rescheduling messages for a random time in the future, and retransmitting them is known as CSMA/CD or Carrier Sense Multiple Access with Collision Detection.

4.3 What Is the TCP/IP Protocol Suite?

If the destination address of a message matches the 48-bit address of your computer, the MAC software passes the message to the TCP/IP protocol suite. The TCP/IP protocol suite is a series of software programs that successively peel and process data packets. Each software program processes the data field remaining from being processed by the previous software layer. This is illustrated in Figure 21.

Protocols in this suite work together by passing their messages up and down the protocol stack. For example, the TCP Protocol (described later in this chapter) takes application data, embeds it in the data field of a TCP message. It then passes the TCP message to the IP protocol where the entire TCP message becomes the data packet of the IP message. At each layer of the TCP/IP Protocol Suite, the software layer is only concerned with its fields and not the contents provided by previous layers.

The TCP/IP protocol suite is software and is usually a component of your computer's operating system. In Microsoft Windows®, Microsoft includes a TCP/IP protocol suite to process Ethernet messages. In industrial devices, the TCP/IP protocol suite may be included as part of an operating system or may be a separate software component. The number of programs included in the suite is dependent on the vendor but always includes both the Internet Protocol (IP) and Transmission Control Protocol (TCP).

Figure 21 – Data Flow through the hardware, IP Layer and UDP/TCP Layers to Applications

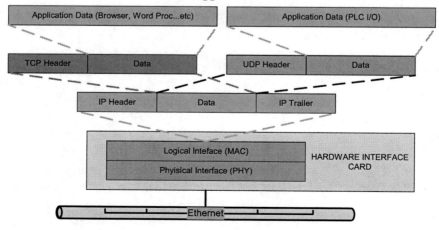

4.4 TCP/IP Protocol Suite – IP Protocol

The IP part of the Internet Protocol Suite is the "Internet Protocol". It is used for almost all Internet communication. When a host sends a packet, it figures out how to get the packet to its destination; when receiving packets, it figures out where they belong. Because it does not worry about whether packets get to where they are going nor whether they arrive in the order sent, its job is greatly simplified. If a packet arrives with any problems (e.g., corruption), IP silently discards it.

In addition to inter-network routing, IP provides error reporting and fragmentation and reassembly of packets for transmission over networks with different maximum data unit sizes. IP provides several services:

- **Addressing.** IP headers contain 32-bit addresses, which identify the sending and receiving hosts. These addresses are used by intermediate routers to select a path through the network for the packet.

- **Fragmentation.** IP packets may be split, or fragmented, into smaller packets. This permits a large packet to travel across a network, which can only handle smaller packets. IP fragments and reassembles packets transparently.

- **Packet timeouts.** Each IP packet contains a Time To Live (TTL) field, which is decremented every time a router handles the packet. If TTL reaches zero, the packet is discarded, preventing packets from running in circles forever and flooding a network.

- **Type of Service.** IP supports traffic prioritization by allowing packets to be labeled with an abstract type of service.

- **Options.** IP provides several optional features, allowing a packet's sender to set requirements on the path it takes through the network (source routing), trace the route a packet takes (record route), and label packets with security features.

The IP Protocol makes great use of another protocol, the Address Resolution Protocol (ARP), to identify the specific destination of a message. When the IP protocol has a message to send, it checks its internal records to determine if it has the physical MAC address for the destination IP address. If not, it issues an ARP Request to obtain the physical address. When the node with that IP address responds, it reports the MAC address that the IP protocol then uses to send the original message.

In addition to ARP requests that discover MAC addresses, there is also a "Reverse ARP" request, which discovers the IP address of a MAC address.

An interesting fact regarding the operation of the IP software is that it is "stateless" and operates silently. Stateless means that it does not remember or care about any previous operation. Unlike almost every other protocol in the TCP/IP Protocol Suite, each message is totally independent of all previous messages. Silent operation means that any invalidly formed packets or packets that can not be verified are simply discarded. No effort is made by the IP Protocol software to inform the protocol layer above it that the message it encoded and sent never arrived.

Table 4-1 — IP Data Packet

Version	Internet header length	Type of service	Total packet length
4 bits	4 bits	8 bits	16 bits
Identifier (a pseudo random tracking number)		Flags	Fragment offset
16 bits		3 bits	13 bits
Time to Live counter (255 max)	Protocol residing above IP		Checksum header
8 bits	8 bits		16 bits
Source IP address 32 bits			
Destination IP address 32 bits			
Padding and options			
Data (total packet length dictated by physical media; for Ethernet, 1476 bytes)			

4.4.1 Why IP Addresses Are Necessary

We discussed earlier that an Ethernet MAC address specifies a particular piece of hardware or a card in a computer. An IP address, on the other hand, is more portable and designates a "virtual" entity. A single IP address can represent one computer or a whole network of computers. Its virtual nature makes it portable and re-assignable. Simply, a

MAC address is like your name or social security number while an IP address is like your mailing address.

Ethernet nodes already have a MAC address; the MAC address could be compared with my Vehicle Identification Number (VIN), which designates a unique piece of hardware. Now if all you had was my VIN number, it might still be difficult to find *me*. Someone else could be driving my car; I might have several cars myself. An IP address is like my mailing address, which is more helpful because it points you directly to me.

The Address Resolution Protocol (ARP) is the functional equivalent of calling the Department of Motor Vehicles and matching my VIN number to my postal address. It obtains the MAC address when the IP address is known. Reverse Address Resolution Protocol (RARP) is the exact opposite: Given a MAC address, it tells you the IP address. These two protocols allow the IP software to build the table of IP addresses and MAC addresses that are so necessary to routing your message to its destination.

IPv4 addresses, the kind we now use, are 32 bits long but writing the individual bits out is tedious. Instead, IP addresses are split into 4 separate bytes, each having 8 bits, represented by dotted decimals. For example:

$$204.101.19.6 = 11001100\ 01100101\ 00010011\ 00000110$$

IPv4 addresses are in short supply, and a new IP protocol, version 6, will eventually replace IPv4.

IP addresses are distributed by the Internet Assigned Numbers Authority (IANA); in turn, assignment of IP addresses is administrated by three sub-organizations: in Asia, www.apnic.net; in the Americas, www.arin.net; and in Europe, www.ripe.net. Internet Service Providers further sub-distribute IP addresses to end users.

4.4.2 The New Internet Protocol Version 6

IPV6 is an expansion of IPV4 and is designed to overcome a number of v.4 limitations. The most obvious improvement is the change from 32-bit IP addresses to 128-bit. This is necessary because the world is literally running out of IP addresses, which are carefully allocated. Other features:

- More levels of addressing hierarchy

- Easier autoconfiguration of addresses

- Quality of service: Messages can be designated for high-priority routing through a LAN or the Internet

- Control over the route through which packets flow through a network via any casting

- Unicast (one-to-one) and multicast (one-to-many) messaging

IPV6 is only beginning to be implemented and will take some time to catch on.

4.4.3 Network ID vs. Host ID

IP addresses are split into two parts: the NetID (a global designation, indicating a specific network somewhere on the Internet) and the HostID (a local address within a network, designating a specific machine). Example: in IP address 196.101.101.4, the HostID is 4 just as you read it. If the last digit is 0, it refers to "this network" and the NetID is 196.101.101.0.

Table 4-2 – Drawing a Line Between the Network and the Host

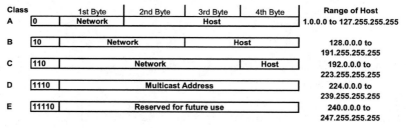

Class	1st Byte	2nd Byte	3rd Byte	4th Byte	Range of Host
A	0 Network	Host			1.0.0.0 to 127.255.255.255
B	10 Network		Host		128.0.0.0 to 191.255.255.255
C	110 Network			Host	192.0.0.0 to 223.255.255.255
D	1110 Multicast Address				224.0.0.0 to 239.255.255.255
E	11110 Reserved for future use				240.0.0.0 to 247.255.255.255

4.4.4 Legacy Address Classes

To facilitate the parsing of IP addresses, five classes were defined long ago, ranging from a few networks with many hosts, to many networks with few hosts, as well as multicasting.

4.4.5 Today: Classless Subnet Masks

Class A, B, and C designations were used until it became apparent that large blocks of IP addresses were being wasted. Whenever a block of IP

Table 4-3 – IP Address Classes, Networks, and Hosts

Class (designated by header bits)	Standard notation	NetID starts at bit #	NetID length (bits)	Number of possible networks	HostID starts at bit #	HostID length (bits)	Number of possible hosts per network	Netmask
A (0)	1.x.y.z to 126.x.y.z	1	7	126	8	24	16777216	255.0.0.0
B (10)	128.x.y.z to 191.x.y.z	2	14	16384	16	16	65534	255.255.0.0
C (110)	192.x.y.z to 223.x.y.z	3	21	2097152	24	8	254	255.255.255.0
D (1110)	Multicast address							
E (11110)	Reserved for future use							

Notes:

1. Address classes are a "legacy" system, which must be explained and understood. *However, today newly issued IP addresses are classless.*

2. 0.x.y.z is not allowed.

3. 127.x.y.z is reserved for loop-back testing, a simple self-test for proper configuration.

4. HostID = 0000000.... (all zeros) means "this network".

5. HostID = 1111111.... (all ones) means "All hosts on this network".

6. A subnet mask is a number which strips the HostID off of an address using the .AND. operation, that is, IP Address .AND. subnet mask = IP address with NetID only. This makes it easy to determine whether an incoming packet is destined for a particular local network. You must know the subnet mask for a given IP address to separate the HostID from NetID.

7. The "A" class theoretically also includes 127.x.y.z, so number of machines is 1677214. A handy subnet calculator is available on the Web: http://www.tcpipprimer.com/subnet.cfm

addresses is issued today, it is issued with a matching subnet mask. Specific notation is used for this:

An address of 203.14.4.13 with a mask of 11111111 11111111 11111111 11100000 (27 1's and 5 0's, or 255.255.255.224) is said to have a *prefix* of 27 and is written as 203.14.4.13/27.

4.4.6 Assigning IP Addresses: Will Your Private LAN be Connected to the Internet?

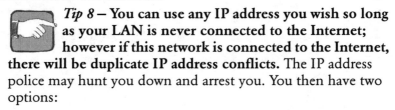 ***Tip 8*** – **You can use any IP address you wish so long as your LAN is never connected to the Internet; however if this network is connected to the Internet, there will be duplicate IP address conflicts.** The IP address police may hunt you down and arrest you. You then have two options:

1. Obtain unique IP addresses from your Internet Service Provider, or

2. Use IP addresses that are *reserved* for private networks. Packets with these address ranges are not forwarded by Internet routers:

Table 4-4 – Reserved Addresses for Private Networks

Class	IP address range	Number of possible combinations
A	10.0.0.0–10.255.255.255	16777216
B	172.16.0.0–172.31.255.255	65536
C	192.168.0.0–192.168.255.255	65536

 Tip 9 – **If you use reserved IP addresses, then the gateway [router] between the LAN and the Internet must be configured as a proxy server to forward Internet packets to each reserved-address device.** Firewalls using Network Address Translation can also do the trick.

 Tip 10 – **Do not confuse the *reserved address classes* A, B and C here with the *general* IP address classes A, B and C, which were discussed a few pages ago.**

4.4.7 Reducing the Number of Addresses Routers Must Advertise with "Supermasks"

The speed with which you can locate sites on the Internet is because thousands of routers have "learned" where various IP addresses can be found. On the Internet, routers must "advertise" to other routers which IP addresses they serve. A mechanism has been devised by which a router can advertise large blocks of IP addresses with a single designation instead of thousands of separate addresses. This is called *Classless Inter-Domain Routing* (CIDR).

A subnet mask concerns the bits that create the NetID. A CIDR is different kind of mask; it concerns the bits in the IP address that are common to all hosts served by that router. So for a block of IP addresses, instead of advertising 2^N addresses, the router must advertise one address, one subnet mask, and one CIDR mask.

Example: A router serves 8 subnets, each containing 256 IP addresses.

Route advertisement: 161.200.0.0

Subnet mask: 255.255.255.0 or 11111111 11111111 11111111 00000000

CIDR mask: 255.255.248.0 or 11111111 11111111 11111000 00000000

Table 4-5 — A Router Serves 8 Subnets, Each Containing 256 IP Addresses

Subnet	Range of IP addresses
1	161.200.0.1-254
2	161.200.1.1-254
3	161.200.2.1-254
4	161.200.3.1-254
5	161.200.4.1-254
6	161.200.5.1-254
7	161.200.6.1-254
8	161.200.7.1-254

Without CIDR, this router would have to advertise 2048 IP addresses. With CIDR, only the route advertisement, the subnet, and the CIDR must be advertised by the router and stored by other routers.

A *default gateway* is an entry in a network configuration table that tells a device where to send a packet if the destination is outside the sender's subnet. The default gateway must be on the same subnet as the sender of the packet.

4.5 TCP/IP Protocol Suite – TCP Protocol

Figure 22 – Handshakes in opening and closing a TCP connection.

Hello would like to open connection

Lets Do that

A Connection is now established

So here is beginning of Data #1

Data #1 received intact

This is last piece of Data # 1

Last Piece Received intact

I am going to hangup now

GoodBye

A highly simplified description of TCP protocol. To send one piece of Data required 7 data exchanges.

Here, Station A opens a connection with Station B, sends the data and closes.

TCP or Transmission Control Protocol is a protocol that *reliably* transfers messages between two computers. Where the IP protocol is concerned only about moving an Ethernet packet to the next node, TCP is concerned with providing a guarantee to the protocol layers above it so that it can with absolute certainty move data between one specific computer and another computer. In postal terms, this is "guaranteed receipt" mail. An indication is returned to show that the message arrived completely and correctly at the requested destination.

TCP uses an inter-network to provide end-to-end, reliable, connection-oriented packet delivery. It does this by sequencing transmitted bytes with a forwarding acknowledgment number that indicates to the receiver the next byte the source expects to receive. Packets not acknowledged within a specified time period are retransmitted. This mechanism allows devices to deal with lost, delayed, duplicate, or mis-read packets.

ETHERNET PROTOCOL & ADDRESSING

TCP also offers efficient flow control, full-duplex operation (sending and receiving at the same time) and multiplexing. Multiplexing provides the capability to transfer numerous message streams over a single connection.

TCP provides the following set of services:

Stream Data Transfer. From the applications point of view, TCP moves *a continuous stream of bytes* through the network or the Internet. The application does not have to parse the data. TCP groups the bytes in *segments*, which are passed to the IP layer for transmission to the destination. TCP segments the data according to its own priorities.

Push Function. Sometimes the application must guarantee that the data reaches its destination. So it pushes all remaining TCP segments in the queue to the destination host.

Close Connection. Similarly, the application pushes the remaining data to the destination.

Reliability. TCP assigns a sequence number to each byte and expects an acknowledgment (ACK) from the receiving station. If the ACK is not received within the timeout period, the data is sent again. Only the sequence number of the first data byte in the segment needs to be sent to the destination. There is no guarantee that packets will arrive in the exact order they were sent, so the receiving TCP puts the segments back in order on the basis of sequence numbers, and eliminates duplicate segments.

Flow Control. When acknowledging receipt of a packet, the receiver also tells the sender how many more bytes it can receive without causing an overflow. This is designated by the highest sequence number it can receive without problems. This is also referred to as a *window*-mechanism.

Multiplexing. Multiplexing is accomplished through the use of ports, just as with UDP protocol.

Logical Connections. Reliability and flow control require TCP to initialize and maintain unique status information for each "conversation." The sockets, sequence numbers, and window sizes for this conversation are called a *logical connection.* Every connection is identified by the unique pair of sockets used by the sending and receiving processes.

Full Duplex. TCP can handle simultaneous data streams in both directions.

When you download large files from the Internet, you will observe some of these characteristics as connections are established.

- Downloads speed up and slow down with variations in network traffic.

- The server you are accessing adjusts its transmission rates to your connection speed.

- When you visit a new Web page, pictures and graphics appear in sequence as new sockets are opened and closed.

- With each mouse click, data moves in both directions.

4.6 TCP/IP Protocol Suite – UDP Protocol

As an alternative to TCP, an application can choose to transmit a message to a destination using UDP (User Datagram Protocol). Unlike TCP, UDP is connection-less and does not provide an acknowledgement of receipt. UDP is the equivalent of first-class mail. Once you drop a letter in the mailbox, there is no record of transmission and no receipt of delivery.

UDP also does not guarantee message order. Ten messages sent over a UDP connection can arrive in an order different than the order transmitted.

UDP has much less overhead than TCP and can be thought of as "faster," in particular because it does not require acknowledgments. UDP is generally used for real-time applications like online streaming and gaming, where missing packets need not be re-sent. In these applications, damaged or missing packets are ignored as more recent data quickly replaces it. For example, if I am mailing you a daily letter describing how many tomatoes I have picked this season, there is no harm if one letter is lost. The next letter arrives with more recent information on the total harvest.

UDP is also employed where upper-layer protocols handle flow control and data stream checking and correcting, such as Netware and Microsoft Networking.

Common applications for UDP include:

- Simple Network Management Protocol (SNMP)
- Domain Name System (DNS)
- Trivial File Transfer Protocol (TFTP)
- Remote Procedure Call (RPC), which is used by the Network File System (NFS)
- Network Computing System (NCS)

4.7 Ports – How the TCP/IP Suite Is Shared Between Applications

The entire suite of TCP/IP protocols is shared between all the applications on your computer. You may have one window searching the Internet, another window running a spreadsheet on a remote server and another window monitoring the stock market. Ports—virtual circuits for the transfer of specific data—are used to accomplish sharing of a TCP/IP stack.

TCP/IP defines 65,536 (2^{16}) available ports which are either common to all Ethernet users, registered to specific applications or unassigned. Each standard service is assigned a port number. Here is a list of the most common services and their port numbers.

Port	Service	Purpose
21	FTP	FTP: File Transfer Protocol, great for downloading programs and files
23	Telnet	Allows remote configuration of a PC or smart device
25	SMTP	Used when *sending* e-mail messages to a mail server
80	HTTP	Used to retrieve Web pages
110	POP3	Used when *receiving* e-mail messages from a mail server
139	NETBIOS	Used for file sharing in Microsoft networking
443	HTTPS	Used to retrieve secure Web pages

A *port scanner* is a software program that probes your computer to detect open ports, which make the system vulnerable to security problems via unwanted external applications.

4.8 Other TCP/IP Application Layer Protocols

DHCP. DHCP stands for Dynamic Host Configuration Protocol and is a clever mechanism for temporarily, automatically assigning IP addresses in a network. It is used quite often.

A typical example of DHCP: You take your notebook computer on a customer visit, and they give you an desk to work at with an Ethernet cable. You configure your network manager in Windows to "Obtain IP address automatically" and every time you boot up your machine, the local router gives it a temporary IP address. Now you can access the Internet and possibly share files with other PC's on that network every time you boot up. This is far easier than manually choosing an IP address (which someone else may have inadvertently taken) every time you go somewhere. Many LANs use DHCP for all of their devices, simply for convenience.

When TCP/IP starts on a DHCP enabled host, it sends a message requesting an IP address and subnet mask from a DHCP server. This server checks its internal database then offers the requested information. It can also respond with a default gateway address, DNS address(es) or NetBIOS Name Server.

When the offer is accepted, it is given to the client for a specified period of time, called a lease. This process can fail if the DHCP server runs out of IP addresses.

SNMP. SNMP stands for Simple Network Management Protocol, and it allows monitoring and managing of a network. A device (automation product, PC, router, switch, or hub) must be enabled with an SNMP agent. The agent stores all variables related to its operation in a database called the Management Information Base (MIB).

The MIB defines all kinds of significant events (reboots, crashes, uplinks, downlinks) and reports such events.

TFTP. Trivial File Transfer Protocol (TFTP) is an extremely simple protocol to transfer files. It is implemented on UDP and lacks most of the features of FTP. It can read or write a file from or to a server. It is not secure and has no provisions for user authentication.

DNS. Domain Name System (DNS) is the convention used on the Internet to point domain names (www.yourcompany.com) to IP addresses.

Since domain names are not likely to be very important in industrial networks, DNS will not be described in detail. However, its basic operation is similar in principle to Address Resolution Protocol (see ARP).

HTTP. HTTP stands for HyperText Transfer Protocol, a TCP/IP protocol that enables the distribution of hypertext documents on intranets and the Internet. Just as with all other TCP/IP application layer protocols such as FTP and SMTP, HTTP is a client/server protocol.

The terms *HTTP server* and *Web server* are somewhat interchangeable, although most Web servers do far more than just HTTP. Similarly, *HTTP client* and *Web browser* are roughly interchangeable, though most browsers do far more than just HTTP.

The web SERVER is the APPLICATION, sitting "above" the stack. HTTP is the APPLICATION LAYER PROTOCOL in the upper layer of the stack, through which the web server access the stack in order to communicate with the CLIENT (like Internet Explorer or Netscape) on the remote machine.

FTP. FTP stands for File Transfer Protocol. It is very popular for moving files between computers. In an FTP session, two connections are opened. One is called a control connection; the other is a data connection; and both use TCP. Each connection can have a different quality of service. Data transfer is always initiated by the client, but either the client or server can be the actual sender.

A popular freeware FTP utility is called WS_FTP and can be downloaded from many Web sites. It conveniently moves, copies, deletes, and renames files between the local computer and a remote host.

Other popular FTP utilities: Coffee Cup FTP, Cute FTP.

Telnet. The Telnet utility provides a standard interface by which a client program on one host may access the resources of server host as though the client were a local terminal connected directly to the server.

Figure 23 – WS_FTP is a popular file transfer utility.

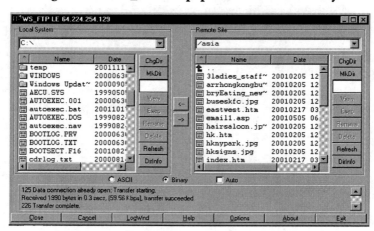

A user on an Ethernet-connected workstation can talk to a host attached to the same network as though the workstation were a terminal attached directly to the host. Telnet can be used across WANs, the Internet, and LANs. Telnet allows the LAN-attached user to log in the same way as the local terminal user. Most Telnet implementations do not include graphics features.

Telnet incorporates three concepts:

* A *Network Virtual Terminal* (NVT) is an imaginary device that applies a common data structure. Every host matches its own terminal characteristics to those of an NVT and expects every other host to do likewise.

* Telnet sessions use the same handshakes in both directions.

* Some hosts have more services than those supported by the NVT, so terminal features can be added and subtracted. Options may be negotiated, so client and server use a set of DO / DON'T / WILL / WON'T conventions to establish the characteristics of their Telnet session.

The two hosts begin by supporting a minimum level of NVT features. Then they negotiate to extend the capabilities of the NVT, according to the actual functions of the real hardware in use. Because of the symmetry in the Telnet sessions, both server and the client may add options.

ETHERNET PROTOCOL & ADDRESSING

Tip 11 – Many switches, "smart devices," and "Internet appliances," especially low-cost simple ones, use Telnet commands for configuration.

4.9 Popular TCP/IP Utilities

PING. Packet InterNet Groper (PING) is the simplest TCP/IP utility, and one of the most useful utilities. It sends one or more IP packets to a destination host, requesting a reply and measuring the round-trip time.

 Tip 12 – The first test you should use to find out if a device is on a network is to attempt to ping it. Normally if you can ping a host, then other applications like FTP and Telnet can also communicate with it. However with firewalls and other security mechanisms, which permit access to networks on the basis of application protocol and/or port number, this may not be possible.

How to use command line PING:

C:\> ping xxx.xxx.xxx.xxx (IP address) <enter>

or

C:\> ping www.nameofthesite.com <enter>

 Tip 13 – Especially when reading PING results on the Internet, remember that it is normal to have occasional packet loss. Also, PING times vary greatly and there is no single figure that designates a "problem" vs. "no problem." On average, the larger your bandwidth, the lower the PING time results. If you experience lousy PING results, the problem is related to the server being down or too busy to reply, or a router between you and the server is down or slow. You can use traceroute to test for this.

Here is an example of the PING utility executed from a DOS Window:

**Figure 24 – PING is a very useful utility, executed from a
DOS command prompt.**

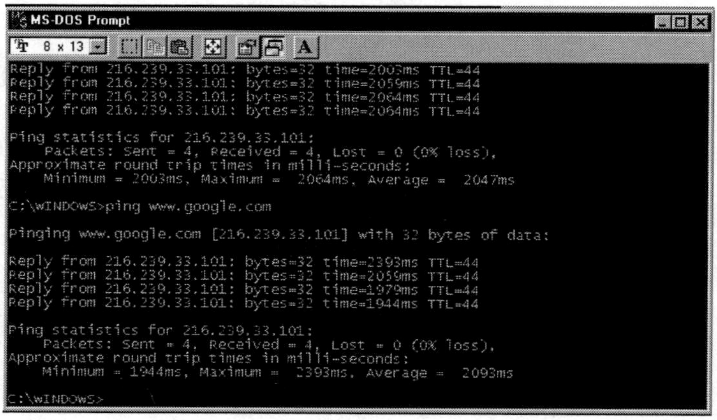

The first thing PING does is translate the URL into an IP address. It is then "pinged" with a packet of information 32 bytes long. Following that is the reply from the server.

time = the total response time for that particular packet.

TTL = Time To Live... which is the number of times this packet is allowed to be retransmitted by routers before being discarded. Each router that handles a packet subtracts one from this value. If TTL reaches zero, the packet has expired and is discarded.

Syntax: This is for DOS; the exact syntax varies with the OS:

ping [-t] [-a] [-n count] [-l size] [-f] [-i TTL] [-v TOS] [-r count] [-s count] [[-j host-list] | [-k host-list]] [-w timeout] destination-list

Options:
-t Ping the specified host until stopped.
To see statistics and continue - type Control-Break;
To stop - type Control-C.
-a Resolve addresses to hostnames.
-n count Number of echo requests to send.
-l size Send buffer size.
-f Set Don't Fragment flag in packet.

-i TTL Time To Live.

-v TOS Type Of Service.

-r count Record route for count hops.

-s count Timestamp for count hops.

-j host-list Loose source route along host-list.

-k host-list Strict source route along host-list.

-w timeout Timeout in milliseconds to wait for each reply.

Traceroute. Traceroute is a TCP/IP network utility that tells you the route from your computer, through each gateway computer [router] at every hop through a LAN or the Internet, to a specified destination computer. The target computer is specified via its Web or IP address. Traceroute also displays the amount of time each hop took. Traceroute is handy for understanding where problems are in a complex network. It also reveals interesting information about the Internet itself.

Syntax: C:\> tracert [IP address or URL]

Figure 25 — Traceroute tells you exactly where a packet goes on its way to its destination.

Netstat. Netstat asks TCP/IP the network status of the local host. Netstat reports:

• Active TCP connections at the local host.

• The state of all TCP/IP servers on this local host and their sockets.

• All devices and links being used by TCP/IP.

• The IP routing tables in use.

ARP. The ARP utility is especially useful for resolving duplicate IP addresses. A workstation is assigned an IP address from a DHCP (Dynamic Host Configuration Protocol) server, but accidentally gets the same address as another workstation. You ping it and get no response. Your PC is attempting to find the MAC address, but it can't because two machines think they have the same IP address.

To resolve this problem, use the ARP utility to view your local ARP table and check which TCP/IP address is resolved to which MAC address.

The ARP protocol belongs to TCP/IP, translating TCP/IP addresses to MAC (media access control) addresses with broadcasts.

In Windows, this table is stored in memory so that it doesn't have to do ARP lookups for frequently used TCP/IP addresses of servers and default gateways. Entries contain not the IP address, the MAC address, and measurement of how long each entry stays in the ARP table.

The ARP table has two kinds of entries, static and dynamic.

Windows creates dynamic entries whenever the TCP/IP stack makes an ARP request and can't find the MAC address in the ARP table. The ARP request is broadcast on the local segment. When the MAC address of the requested IP address is found, Windows adds that information to the table.

Static entries work the same way as dynamic, but you have to manually implement them using the ARP utility.

The ARP Utility. To use the ARP utility in Windows 95/98, follow these steps:

1.   Open the DOS window.

2.   At the command prompt, type **ARP** plus any switches you need:

ARP -s inet_addr eth_adr [if_addr]
ARP -d inet_addr [if_addr]
ARP -a [inet_addr] [-N if_addr]

Table 4-6 – The ARP Utility

-a	Displays present ARP entries by examining the current protocol data. If inet_addr is specified, the IP and Physical addresses for only the specified computer are shown. If more than one network interface uses ARP, entries for each ARP table are shown.
-g	Same as –a
inet_addr	Specifies an internet address.
-N if addr	Displays the ARP entries for the network interface specified by if_addr.
-d	Deletes the host specified by inet_addr.
-s	Adds the host and associates the Internet address inet_addr with the Physical address eth_addr. The Physical address is given as 6 hexadecimal bytes separated by hyphens. The entry is permanent.
eth_addr	Specifies a physical address
if_addr	If present, this specifies the Internet address of the interface whose address translation table should be modified. If not present, the first applicable interface will be used.

5.0—Basic Ethernet Building Blocks

Even a single Ethernet network can be quite extensive, with up to 1024 nodes, hundreds of cables, and infinite possible combinations of hubs, switches, bridges, routers, network interface cards, and servers. This chapter describes these devices and their functions.

To understand the devices in this section the concept of a *collision domain* must be understood. Whenever two or more devices on an Ethernet segment begin transmitting at the same instant, there is a collision and neither message is transmitted. More and more collisions occur as the number of devices on a single segment increase until none of the messages can be delivered. Limiting the number of devices on a segment – the number of devices in a *collision domain* – solves this problem.

5.1 Devices

Hubs

Hubs are the simplest method of redistributing data on Ethernet. Hubs are "dumb," meaning that they do not interpret or sort messages that pass through them. A hub can be as simple as an electrical buffer with simple noise filtering; it isolates the impedances of multiple "spokes" in a star topology. Some hubs also have limited "store and forward" capability. In any case, hubs indiscriminately transmit data to all other devices connected to the hub. All of those devices are still on the same collision domain.

Note: Hubs are not assigned MAC addresses or IP addresses.

Figure 26 – Ethernet hub.

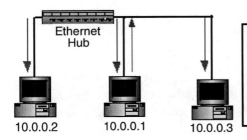

Here, when PC (10.0.0.1) sends a message to 10.0.0.2, it is received by all the PCs and no other PC communicate til 1 is done with the transmission fully.

Workgroup Hubs. Workgroup hubs are usually stand-alone units with four to eight ports. *Stackable hubs* often have many more ports and can be linked together to form a "super hub," which links even more devices to the same collision domain.

Segmented Hubs. Segmented hubs allow you to divide the available ports among multiple groups and collision domains. Each group you define is isolated from the others, as though you were using completely separate hubs. Bridges allow communication between segments.

Two-Speed Hubs. Two-speed hubs allow multiple baud rates to operate on the same hub, auto-detecting the data rate at each port and linking ports together with a speed matching bridge.

Managed Hubs. Managed hubs have modest levels of intelligence and can be controlled remotely via a configuration port. This allows ports to be turned on and off, segments to be defined, and traffic to be monitored.

Repeaters. Repeaters are essentially two port hubs. They simply clean up the signal and boost the signal level for large distances.

Stacking or "Crossover" Cables. Stacking or "crossover" cables allow multiple hubs to be daisy-chained. You can't use standard cables to link two hubs together (or two NIC cards together) because they will link the transmit pins to transmit pins instead of connecting transmit pins to the receive pins. Some hubs have *crossover ports,* which allow standard cables to be used.

Bridges

Bridges allow traffic to selectively pass between two segments of a network. Bridges operate at layer 2 of the OSI model and effectively extend the reach of each segment. Bridges make their forwarding decisions on the basis of the MAC address.

 Tip 14 — You should use a bridge when your network traffic can be clustered into "devices on segment A that mostly talk just to each other" and "devices on segment B that mostly talk just to each other." The bridge handles the cases where devices on A must talk to B, and the rest of the time it reduces the traffic on each side.

Figure 27 – Bridge.

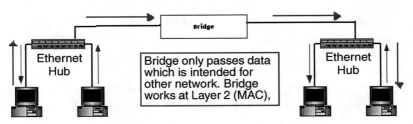

Intelligent Bridges. Intelligent bridges learn over time what devices are connected on each side and "figure out" which messages to forward and which ones to block. Such a bridge will automatically adapt to changes made to the networks over time.

Tip 15 – **Care should be taken not to form "loops" on networks with multiple connecting bridges.** The IEEE 802.1 "spanning tree algorithm" removes loops. One bridge in a loop becomes the "root" and all other bridges and sends frames toward the root bridge.

Note: Bridges are not assigned MAC addresses or IP addresses.

Switches

A switch is an intelligent bridge with many ports. A switch learns what addresses are connected to each port and sends messages only to their intended destinations.

There are two basic types: *Cut through switches* forward a message to their destination as soon as they recognize the intended address. *Store and forward switches* hold the packet in memory and examine the entire contents of the packet first. This enables it to trap errors and prevent those bad packets from being sent through the network. A store and forward switch can also hold the packet until traffic on that segment disappears, reducing collisions.

Switches usually operate at layer 2 (switching decisions based on Ethernet MAC address) and some operate on layer 3 (switching decisions based on IP address). Layer 3 switches can be used in place of routers.

Full-duplex switches handle both transmit and receive lines simultaneously.

Multispeed switches handle segments with multiple data rates. Typically these have several "normal speed" (e.g, 10BASE-T) ports and one high-speed (e.g., 100BASE-T) port. Several PCs can connect via the 10-M ports and the switch can relay all traffic to a server or backbone via the 100-M port with no "bottleneck" from aggregating the data.

Many industrial applications require some level of determinism, and Ethernet is often criticized for its lack of determinism. Switches make deterministic performance possible by eliminating collisions. IEEE802.1p allows prioritization of messages at layer 2 with a 16-bit additional header. Rigid priorities on the network allow important messages to be sent without collisions.

Note: Switches are not assigned MAC addresses or IP addresses.

Routers

A router's job is to forward packets to their destination, using the most direct available path. Routers make their decisions on the basis of IP addresses. When a packet comes in, a lookup table determines which segment it should be routed to.

Many times the segment has only another router instead of a final destination. The implication is that the intended destination is remote. The IP address of the next router (there could be many) is called the *default gateway.*

Routers maintain tables of IP addresses on each segment and "learn" the most direct paths for sending data. When the network changes, it takes time for routers to accommodate the new information. On the Internet, if you have a Web site www.yourwebsite.com, the URL points to the IP address of the host server. If you change hosts, you get a new IP address and register the change with the Domain Name Server. It may take several days for DNS servers across the Internet to update their tables and point you to the new IP address.

Routers are protocol-dependent because they operate at layer 3. A router that handles TCP/IP may not be able to work with Novell SPX/IPX.

Types of Routers

2-Port Routers. 2-port routers link only two networks.

Multi-Port Routers. Multi-port routers link several networks.

Access Routers. Access routers use modems (e.g., ISDN, v.34, v.90) to access the Internet, usually on-demand.

Bridging Router (Brouter). Bridging routers, or brouters, change from being routers to being bridges when they receive a packet they don't understand. They just go ahead and send the message.

Terminal Servers. Terminal servers connect multiple serial devices (RS-232, -422, -485) to Ethernet. The term comes from terminals on mainframes, connected to the LAN via serial port.

Figure 28 – Types of routers.

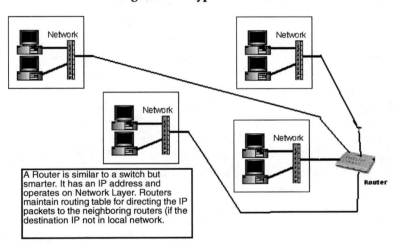

Thin Servers. Thin servers link a single device to Ethernet and allow COM ports on the other side of the network to appear as though they are local to your PC, even though they may be on the other side of the world.

Network Time Servers. Network time servers use a Global Positioning System (GPS) to provide accurate local time for synchronization of devices and time stamping of events.

Gateways

Gateways convert messages from one protocol to another. Examples could include Modbus on RS-232 to Ethernet Modbus/TCP, or

DeviceNet to EtherNet/IP. In most cases, the physical layers, protocols, and speeds are all different.

 Tip 17 – **Gateways normally require configuration to work properly and are normally thought of as "band aids" rather than permanent or global solutions.**

Interface Cards

A NIC links your PC to Ethernet via the PCI, ISA, PCMCIA, PC/104, or other buses. NICs handle layers 1 and 2, while the host processor in the PC handles everything else.

5.2 Determinism, Repeatability, and Knowing if It's "Fast Enough"

Many people confuse determinism with speed. Some definition of terms is helpful:

Deterministic means that a system is guaranteed to respond within a designated period of time: less, but no more. The term really is not very meaningful unless a time specification is included with it.

Example: "This PLC is deterministic to 10 milliseconds" means that when an input state changes, the corresponding output state change will occur no more than 10 ms later.

The term *repeatable* defines the space between the lower *and* upper limit for response time. A spec for repeatability designates the width of a time window.

Example: "This PLC is repeatable within 2 milliseconds" means that the response time will never vary by more than ±1 ms. This statement does not specify the response time though, only the *consistency* of the response.

 Tip 16 – **A full understanding of a system's response capabilities normally requires specification of both determinism and repeatability.** Some automated processes require determinism and repeatability; some require only determinism; some require one or both only "most of the time," and some require neither.

Your watch is both deterministic and repeatable, within microseconds. An E-Stop must absolutely be deterministic, but repeatability per se is not usually so important: there's no good reason to have a fixed amount of delay before the system shuts down. A pneumatic valve that rejects black grains of rice into a waste bin upon detection by a vision system must be deterministic *and* repeatable, otherwise the puff of air may come too soon, which is just as bad as coming too late.

"Fast enough" is another issue. Most processes do not require absolute determinism; they really require that systems respond within a certain amount of time, most of the time. So you define the requirement as a statistical probability that the system will respond within, say, 10 ms 99.9% of the time.

In those cases, even a nondeterministic collision-based Ethernet system will be acceptable, so long as the network loading is below acceptable limits 99.9% of the time.

Achieving Determinism on Ethernet

Half-duplex Ethernet has inherent collision problems, and though the CSMA/CD protocol offers a good solution, it is inherently nondeterministic. Complex formulas exist for calculating network loading and response-time probabilities, but *if true determinism is necessary, you should isolate collision domains by using switches instead of hubs.*

With switches, the remaining determinism problems are caused by their throughput limitations. If the switch is not able to handle the full speed on each port, or if the number of packets sent to an output port exceeds the bandwidth of that port and fills the output buffer, this causes a non-deterministic buffering delay. Higher protocol layers at the stations must handle lost packets.

The following methods are used to prevent switches from overloading:

- *Flow control:* The switch sends PAUSE packets on a full-duplex port if the number of packets received on the port is more than the switch can handle.

- *Back pressure:* If the traffic load exceeds the switch's capacity, the switch acts like a port operating in half-duplex mode. It makes the transmitter think the collision domain is busy.

- *Priority:* Ethernet packets that are designated as high priority are put in a high-priority queue. Those packets are sent ahead of the low-priority packets, which might possibly be dropped. This is the most "deterministic" approach to the problem.

How Priority Messaging Works

Many switches now support priority, with two or more output queues per port, where the higher priority queue(s) are designated for time-critical data with quality of service. Each vendor uses a different algorithm for selecting the queues, but in general there are two approaches:

- Round-robin: After X packets are unloaded from the high-priority queue, a low-priority packet gets its turn to go out.

- High priority always takes precedence: Low-priority packets are sent only when the high-priority queue is empty.

In any case, a high-priority message can still be delayed by a low-priority message if the low-priority message has already started transmitting when the high-priority message enters the switch. But it's not that difficult to calculate worst-case scenarios for this.

The worst-case delay for high-priority packets holds, regardless of low priority traffic. Several priority implementations exist with respect to how a packet is identified as a high-priority packet. The priority handling depends on the switch functionality.

How Switches Determine Priority

- *Based on MAC addresses:* Both the MAC source and destination address can be prioritized. The switch must be a "managed" switch so the user can choose high-priority MAC addresses. This is a fairly rigid approach.

- *Based on high-priority physical port:* One or more switch ports can be designated as high priority, so all packets received on these ports are considered high-priority packets. Most switches that work this way are "managed." A *managed switch* is one that is externally configured for optimum performance—this provides extra functionality with the disadvantage of extra complexity.

- Based on priority tagging: IEEE 802.1p and IEEE 802.1q designate an additional Tag Control Info (TCI) field for the Ethernet MAC header. This adds a 3-bit priority field that is used for priority handling, allowing 8 levels of priority. Most priority tagging Ethernet switches have only two or four queues, so the network configuration must account for this limitation. The advantage is that no switch configuration is necessary. Unfortunately most stations today do not support priority tagging. The switch can be configured to remove the tags after switching, but this requires managed switch operation. Another problem could be other switches don't support priority tagging and will not forward the longer packets.

 Caution: Even though they may be rated for 10-, 100-, or 1000-Mbps operation, many switches cannot sustain full traffic loads at their rated speeds. This can create problems on high-speed deterministic systems. If this presents a potential problem, ask your vendor what the throughput of their switch is, and how the switch allocates its time to competing devices.

Drivers and Performance

All other network layers, including TCP/IP, are handled in software. Nearly all PC operating systems, including Windows, Linux, DOS, UNIX, VxWorks, etc. have TCP/IP built in. Software applications for control or operator interfaces have drivers that pass application data, including higher-layer protocols like Modbus/TCP and EtherNet/IP to TCP/IP.

 Tip 18 – There can be significant performance issues and delays with respect to driver and application performance, and there are no defined standards for driver performance. Many drivers simply are not written to serve the needs of deterministic applications.

Response time can vary considerably on the basis of CPU speed, memory, how well the drivers are written, what other applications are running on the PC, etc.

Tip 19 – There can be no doubt that in most cases, the speed of industrial Ethernet networks will be limited by software and drivers, and not by Ethernet itself.

5.3 Cost Issues: Ethernet "Smart Devices" vs. PCs

Between 1995 and 2000, the Internet caught on like wildfire with an adoption rate far exceeding any previous technology. One of the key reasons was that the price of networking via modems and office LANs dropped to the point where connecting to the Web was a "no brainer." Critical mass was achieved, and the rest, as they say, is history.

The next revolution, now underway, is clearly "Internet Appliances"—smart devices that are accessible via WANs, LANs, and the World Wide Web. However, there are important cost issues.

Ethernet NIC cards are fairly simple devices with few circuit board components. A NIC card is a "non-intelligent" device: It is purely a layer 1/layer 2 physical interface between the Ethernet cable and the PC's ISA or PCI bus, and has no microprocessor. Layers 3 and above are all handled by the host processor, most commonly a Pentium-class device. TCP/IP software is included in popular operating systems like Linux, Windows, MAC OS, and UNIX, and the development of this software is quite mature. For these reasons, consumer-grade NIC cards can be purchased for less than $25, a small fraction of the total cost of a PC. The hardware and software integration is standard and fairly easy.

Similarly, a typical 56K modem can be purchased for under $50, and most modems are similar to NIC cards in that they contain little intelligence. The majority of the software, including Digital Signal Processing functions, is done by the host (e.g., Pentium) processor. Its enormous bandwidth, combined with the fact that most desktop applications do not require real-time deterministic performance, means that the extra processing burden can be offloaded to the host.

There is a higher-performance class of modem called a "hardware modem," which has a Digital Signal Processing chip and related software residing on the card itself. The modem itself is an embedded computer, and it is certainly superior to a "WinModem." It is not unusual to pay $100 or more for such a device. Those who have used other industrial networks, such as Profibus, Remote I/O, or Modbus Plus, know that PC cards for these networks, which are analogous to hardware modems, sometimes cost upward of $1000. This is due to the higher parts count, the low volume of the marketplace, complexity of the protocols and tools, and support costs.

The design of an Internet appliance invokes cost issues, illustrated by the difference between WinModems and hardware modems. Adding Internet capability to a temperature controller, PLC, drive, or barcode reader means running a TCP/IP software stack and adding additional hardware. Such devices typically have small 8- and 16-bit processors.

Designers rarely have the luxury of using 1-GHz processors and off-the-shelf software. Unless the volume is in the tens or hundreds of thousands of units, it's easier to add an additional co-processor dedicated to communication tasks than to add TCP/IP software to an already burdened microprocessor.

So Internet-enabling a smart device is comparable to adding a hardware modem to the bill of materials cost. The cost impact to a device whose base price is only a few hundred dollars can be substantial.

The bottom line is: networking a smart device is more expensive than networking a PC. This will slowly change as the market for embedded Ethernet devices evolves; new communication System-on-Chip components and turnkey software stacks for product developers are emerging at lower and lower prices. Nevertheless, given the low production quantities typical of factory automation products, expect to pay more for industrial Ethernet devices.

Tip 20 — **Just because it's Ethernet doesn't make it cheap, and there's no reason why an Ethernet product *has* to be cheap.** Cost justification can easily come from the additional value of additional information that is available: diagnostics, data acquisition, and remote configuration.

6.0–Network Health, Monitoring, & System Maintenance

By Mark Mullins. Reprinted with permission from the **Industrial Ethernet Book,** *www.ethernet.industrial-networking.com, ©2001 GGH Marketing Communications and Fluke Networks.*

6.1 What Is It that Makes a Network Run Well?

Fluke Networks has profiled dozens of networks worldwide in an effort to determine the answer. In our research, the best run-networks had thirty-five times less downtime, resulting in annual savings of over $227,000. Not surprisingly, users of these networks were the most satisfied of all groups studied. One surprising conclusion is that the number of support staff per end-user of these well-run networks was actually lower than that of the poorly-performing networks.

So how does a network support group enter this desirable group? In studying this question, Fluke Networks uncovered seven "best practices" of well-run networks. They are:

- Management Involvement
- Preparation & Planning
- Problem Prevention
- Early Problem Detection
- Quick Problem Isolation and Resolution
- Invest in Tools and Training
- Quality Improvement Approach

Having the right tools for monitoring, documenting, and troubleshooting your network helps with nearly all of these areas. Let's look at each of these three functions and discuss the tools for each.

Monitoring

Monitoring your network is essential to find problems before they become serious, and just to have a general idea of what is going on in your network. When monitoring, there are a number of key questions you'll need the answer to:

- Who's talking to whom?

- What are they talking about?

- Are there problems out there?

The most important tools are protocol analyzers, embedded **RMON** (Remote MONitoring) agents, and external RMON agents.

Protocol analyzers allow the network engineer to capture traffic passing by on the network, and then decode that traffic in order to understand the traffic. For example, single frame my contain IP addressing information, TCP flow control, and HTTP commands. The analyzer needs to be able to decode each of the protocols so that the user is not presented with a bunch of unintelligible hexadecimal. Billions of such frames may travel over a network in the course of a day, so the analyzer needs to be set up to filter, that is to capture only certain types or sources of frames, as well as trigger, or start capturing traffic after a certain type of frame is detected.

Low cost analyzers have trouble keeping up with fast or busy networks, while higher-priced ones offer more memory, and specialized hardware to keep up with even gigabit ethernet. Higher priced ones also decode more protocols, and offer expert analysis of traffic to find problems faster.

Embedded agents are found in most of today's switches and routers, and collect information on activities in each interface of the device. This information is stored in a Management Information Base (MIB), and can be accessed by devices using Simple Network Management Protocol (SNMP). External RMON agents do much the same thing, except that they offer much greater depth of information, and they have to be purchased (where embedded RMON is usually a no-cost feature).

In terms of depth, embedded RMON agents generally offer only a very-high level overview of what's happening on the network interface: utilization statistics (how 'busy' the interface is), and error counts. Some embedded RMON agents can also generate alarms when certain thresholds are exceeded. External agents, on the other hand, generally offer much greater detail, such as which devices are using the port, and the ability to capture traffic, like a protocol analyzer.

Most newer external agents support RMON2, which adds the ability to track application-layer traffic. This is important in determining who is really using bandwidth. A quick comparison of the detail yielded by the three approaches is instructive. An Embedded RMON agent would tell you that Ethernet Interface 42 is very busy. An external RMON probe could tell you that a certain PC is sending a lot of traffic to the router. An RMON2 probe could tell you that 'Bob's PC' is sending HTTP traffic to the Web site 'hotjobs.com. Obviously, RMON2 provides much more detail. And, not surprisingly, that detail comes at a cost.

Monitoring Switched Networks

The main issue with external probes and protocol analyzers is where to put them. Before the advent of switched Ethernet, a probe placed anywhere on the network could see all the traffic on the network and provide total visibility. Today, switched networks provide higher performance and faster response, and are recommended for industrial networks. Unfortunately, switched networks only send traffic to the intended recipient, so special allowances must be made so that the probe can monitor the relevant part of the network.

One of the simplest methods is to use port mirroring. By appropriate commands to the switch, it can be configured to copy traffic at one interface to another one where the probe is connected. The advantages of this method are the fact that it allows you to monitor whichever port you want. The disadvantage is that while monitoring that port, you have no idea what's happening on any other port. In addition, port mirroring will generally forward only good frames, and often, it's the bad ones you're looking for. Finally, if the switch is very busy it may not send all the frames to the mirror port - so that one critical frame causing the problem might be missed.

> *Tip 21* – **For more thorough monitoring, a tap may be installed into critical links.** This is simply a hardware device that allows the probe to see all the traffic on the link. Unlike mirroring, taps never miss a frame or an error. However, they have to be installed on every port that must be monitored. And like mirroring, they give visibility into only one link at a time.

To get complete vision when monitoring a switched network, a combination of approaches must be used. Embedded agents can be monitored to get an overview of what's happening on every interface in the network. External RMON2 agents can be installed with taps to provide constant monitoring of key links, such as those between switches, to servers, and wide area networks. Additional RMON agents can be connected to switches and connected as needed using port mirroring to monitor problem ports found with the embedded agents.

Documenting

Tip 22 – **Documenting your network is essential for two reasons.** First, when it becomes necessary to upgrade or expand your network, you'll need an idea of where to start. Second, knowing the normal state of your network is essential when it's time to troubleshoot. If a doctor didn't know what a normal temperature or blood pressure is, those measurements would be meaningless. Each network has its own normal operating condition, so it's important to know what yours looks like—before a problem arises.

A good example of this is a network at an automotive plant in Michigan, where Fluke Networks was offering training on network documenting. This customer had recently upgraded its network and was extremely pleased with how well it was operating. In the course of the class, we documented a number of interesting characteristics of this network —over 1,100 stations in one collision domain, over 400 errored frames in two hours, sustained peak utilization over 70%, and an average of 2% collisions. Any of these would be considered serious problems in most networks, but this was 'normal' for them. If they ever experience a problem they need only look for what changed, rather than waste time tracking down unrelated issues.

Documenting is the process of recording the state of your network. The two main questions are, "what's out there", and "how is it performing". By keeping records provided through monitoring systems, a good idea of normal performance can be obtained. Some additional documenting tools can be valuable.

The first of these can be called SNMP/Ping Monitors. These discover the devices in the network and then gather SNMP information form these devices. They also offer some powerful documenting features. Microsoft Visio, for example finds the devices and then provides a complete network diagram at a very reasonable price. Fluke Networks' Network Inspector provides a wide variety of reports, and can plot SNMP statistics for up to a 24 hour period. Other reporting packages are available from Concord Communications, InfoVista, and Visual Networks.

Troubleshooting

When something goes wrong, it is usually important that it get fixed as fast as possible. Many of the tools noted above can help with the troubleshooting process. In fact, if everyone installed everything noted above, and never let any users or applications near the network, there wouldn't be much need for specialized troubleshooting tools. For the real world, however, a number of specialized tools are available to solve problems fast.

The most common of these is the Protocol Analyzer, loaded onto a laptop computer. While indepth analysis capabilities allows this to tackle the most challenging of network problems, the complex set-up and limited vision in switched networks make them a troubleshooting tool of last resort.

The next most common tool is the *cable tester*, which can help track down the most common cause of network problems—cabling. Basic and advanced types are available. Basic testers will find broken cables, shorts, and split pairs. A split pair occurs when two channels constituting a transmit plus and transmit minus are not connected to a pair that is twisted together. The result can be transmission errors or even a complete breakdown in communications. Many low cost testers cannot find this problem—use of these is not recommended. Some basic testers can display the distance to the fault, which can greatly speed troubleshooting.

 Tip 23 – **Advanced cable testers not only find these common problems, but can also determine the performance level of the cable.** In the days of 10-Mb Ethernet, almost any cable could handle the requirements. Higher-speed networks, at 100-Mb and gigabit speeds, place significantly higher demands on the cabling and more advanced tools are needed to determine if the cable is up to the task. These advanced testers measure parameters such as crosstalk and return loss and can certify the performance of cabling for high speed networks. They also cost about four times what a basic tester costs! If cable performance is verified with an advanced tester at installation, most sites need only the basic tester for daily troubleshooting. However, as higher performance networks become more common, the need for advanced testers will grow accordingly.

A new class of tester, the Integrated Network Analyzer, offers the fastest approach to network troubleshooting. These devices incorporate the most commonly used capabilities of protocol analyzers and cable testers to provide a complete solution for troubleshooting. Portability and ease of operation an two key features. Some also offer advanced features such as the ability to discover devices on the network (like SNMP/Ping Monitors) and SNMP queries of network devices.

Mark Mullins is Marketing Manager for Enterprise Systems at Fluke Networks in Everett, Washington, USA. He holds a Bachelor's in Computer Science and an MBA from the University of Washington. He has been with Fluke for 21 years, and was one of the founders of Fluke Networks.

6.1 Popular PC-Based Ethernet Utilities, Software, and Tools

PC-based network sniffers operate on a PC that links to an Ethernet LAN via a hub (not a switch) and collect data as it goes by. They break down and organize Ethernet frames and/or TCP/IP packets so you can make sense of what's happening on your network. Use them to identify chattering nodes, corrupted data, and mysterious sources of network traffic.

If your TCP/IP sessions "hang up," a sniffer might tell you which device sent the last packet and which one failed to respond. Similarly, if devices are responding slowly, time stamps will show you which system is waiting and which system is responding slowly. The sniffer can monitor broadcast or multicast storms and packet errors. By recording and displaying the traffic on the Ethernet wire, or a filtered segment of the traffic, you will pinpoint problems and intelligently improve network performance.

Table 6-1 — Sniffer Basic from Network Associates

Package	Sniffer Portable from Network Associates http://www.networkassociates.com/us/products/sniffer/mgmt_analysis/ sniffer_portable.htm
Platform:	Windows
Licensing:	
Description:	Sniffer Portable is a full-featured Performance Management tool designed to maintain, troubleshoot and fine-tune networks. The analyzer captures frames building a database of network objects from traffic and using that to automatically detect network anomalies.

Table 6-2 — EtherPeek by the AG Group

Package	No change
Platform:	No change
Licensing:	No change
Description:	Network traffic and protocol analyzer. Selected first in 2001 Network World Global Test Alliance analysis.

Table 6-3 — The Gobbler by Tirza van Rijn

Package:	The Gobbler by Tirza van Rijn, University of Delft, The Netherlands http://www.umich.edu/~archive/msdos/communications/wattcp/delft /gobbler.zip
Platform(s):	DOS
User interface:	Text graphics
Licensing:	Freeware with source code available.
Description:	A highly regarded freeware DOS Ethernet sniffer. Decodes the Ethernet, IP, TCP, and UDP layers and low-level protocols like ARP and ICMP. The interface is easy to maneuver.

Table 6-4 — Analyzer, WinDump, and WinPCap

Package:	Analyzer, WinDump, and WinPCap by Piero Viano, Paolo Politano, and Loris Degioanni http://netgroup-serv.polito.it/analyzer/
Platform(s):	Windows
Licensing:	Freeware
Description:	Analyzer is a user interface built on top of WinPCap (http://netgroup-serv.polito.it/winpcap/), which is a Windows port of Libpcap. They have also ported tcpdump (http://tangentsoft.net/wskfaq/resources/#tcpdump) to a Windows program called WinDump (http://netgroup-serv.polito.it/windump /install/). The features and usability of the interface are very good. Documentation is in Italian.

Table 6-5 — Ethereal

Package:	Ethereal http://www.ethereal.com
Platform(s):	UNIX, Windows
Licensing:	GPL
Description:	Ethereal is a free network protocol analyzer for UNIX and Windows. It allows you to examine data from a live network or from a capture file on disk. You can interactively browse the capture data, viewing summary and detail information for each packet. Ethereal has several powerful features, including a rich display filter language and the ability to view the reconstructed stream of a TCP session. Ethereal can remotely debug network problems: use Telnet to access a UNIX server at a remote site, upload a copy of Ethereal, record network traffic and save it to a file, retrieve the file, and view it with Ethereal. It may save you a plane ticket to some remote location.

Table 6-6 – Unico Network Sniffer

Package	Unico Network Sniffer http://www.unico.com.au/sol_np_es.htm
Platform:	Windows
Licensing:	
Description:	Monitors TCP/IP traffic over Ethernet connections. Runs on any workstation with connections to the relevant Ethernet segments. Provides the ability to monitor live traffic in a non-invasive fashion.

Table 6-7 – Ultra Network Sniffer

Package	Ultra Network Sniffer http://www.gjpsoft.com/UltraNetSniffer/
Platform:	Windows
Licensing:	Single Computer
Description:	From GJPSoft, Ultra is a powerful network sniffer, packet sniffer, sockets sniffer and protocol sniffer. It consists of a well-integrated set of functions that you can use to resolve network problems. It will list all network packets in real-time from multinetwork cards (including Modem, ISDN, ADSL) and also support capturing packet base on the application (SOCKET,TDI etc.). Plug-ins for different protocols such as Ethernet, IP, TCP, UDP, PPPOE, HTTP, FTP, WINS, PPP, SMTP, POP3, and so on.

Table 6-8 – CommView

Package	CommView http://www.tamos.com/products/commview/
Platform:	Windows
Licensing:	Single Computer
Description:	CommView lets you see the list of network connections and vital IP statistics and examine individual packets. Packets are decoded down to the lowest layer with full analysis of the most widespread protocols. Full access to raw data is also provided. Captured packets can be saved to log files for future analysis. It is designed for Internet users and small and medium-size networks and can run on any Windows system. CommView features full decoding of an impressive number of protocols including: ARP, BCAST, BGP, BMP, CDP, DAYTIME, DDNS, DHCP, DIAG, DNS, EIGRP, FTP, G.723, GRE, H.225, H.261, H.263, H.323, HTTP, HTTPS, ICMP, ICQ, IGMP, IGRP, IMAP, IPsec, IPv4, IPv6, IPX, HSRP, NCP, NDS, NetBIOS, NFS, NLSP, NNTP, NTP, OSPF, POP3, PPP, PPPoE, RARP, RADIUS, RDP, RIP, RIPX, RMCP, RPC, RSVP, RTP, RTCP, RTSP, SAP, SER, SMB, SMTP, SNA, SNMP, SNTP, SOCKS, SPX, SSH, TCP, TELNET, TFTP, TIME, TLS, UDP, VTP, WAP, WDOG, 802.1Q, 802.1X.

Table 6-9 – Atelier Web Security Port Scanner

Package:	Atelier Web Security Port Scanner by Atelier http://www.atelierweb.com/pscan/
Platform(s):	Windows
Licensing:	Shareware
Description:	A port scanner probes your PC for open ports that may be used by external applications and thus present security threats. It has an Internet Assigned Numbers Authority (IANA) Ports Database, a useful function that tells what purpose the port is registered for, who registered it, and what other uses it's known to have. It also tells you which program is listening on a port. The user interface is very good.

7.0—Installation, Troubleshooting, and Maintenance Tips

7.1 Ethernet Grounding Rules

 Tip 24 – **The shield conductor of each coaxial cable must be grounded at one point only; otherwise you will create ground loops.** On coax, this is often done at the location of a terminator. Many terminators provide screw terminals for this purpose. You should check for exposed wire at other locations since it could make contact with other conductors or ground points.

Ethernet Grounding Rules for Coaxial Cable

• 10BASE-5 (Thick Ethernet): Grounding is a requirement.

• 10BASE-2 (Thin Ethernet): You can ground if your local electrical code requires it.

Grounding coaxial cable is generally good; it dissipates static electricity and makes your network safer. Many local electrical codes require network cables to be grounded at some point.

 Tip 25 – **Many Ethernet segments are not grounded though, and grounding can add complications to an otherwise working network.** But always follow the electrical codes. A segment should be grounded only at one end of the coaxial segment.

 Tip 26 –Do Not Use Copper Cables to Link Buildings! Copper cable attracts lightning strikes.

The ground potential between the two buildings may be different. This can introduce transient voltages and any number of dangerous problems.

 Tip 27 – **Use fiber to connect buildings instead.**

Twisted-Pair-Cable Types

Twisted-pair cabling is categorized as follows:

- Category 1 is mostly used for telephone connections. Do not use for computer networking.

- Category 3 works up to 16 Mbps and may be the most common installed twisted-pair format.

- Category 5 works up to 100 Mbps and is the most popular kind of cable sold by computer vendors.

- Category 5E supports Gigabit Ethernet and is preferable if there is any possibility of upgrading your system to 1000 Mbps.

- Category 6 includes all of the CAT5E parameters but extends the test frequency out to 250 MHz, exceeding current Category 5 requirements. Category 6 will be the most demanding standard for 4-pair UTP terminations based on RJ-45 connectors.

 Tip 28 – **When you use a specific grade of cable, all components and interconnects on the network must also be equal to that quality level.**

Grounding for Shielded Twisted Pair

 Tip 29 – **STP must be grounded because of the shield.** The ground should be connected only at one end. CAT5 STP patch panels normally provide a grounding strip or bar.

Hubs and switches don't provide grounding. If you attempt to establish a ground with an active device and it experiences an electrical disturbance, surges will occur on the cable. This will damage all equipment attached to the LAN and may create a fire hazard.

Reducing Electromagnetic Interference (EMI)

Much time and money can be saved by routing network and power cables through a single raceway. But the cables become highly susceptible to noise coupling. *Common mode voltage* signals, that is, voltage induced equally on all signal conductors by a power line, are sometimes a single-digit percentage of the power-line voltage. Because transformer-coupled systems as used on 10BASE-T and 100BASE-T reject common mode voltages, this is not a severe problem. However, these common mode problems can lead to large differential voltage and corrupt data.

⚠️ *Caution:* *Standard Ethernet magnetics typically have 1500-V isolation ratings. IEC standard 1000-4-6 establishes a common reference for evaluating the performance of industrial-process measurement and control instrumentation when exposed to electric or electromagnetic interference. This standard requires 2000-V surge and fast transient burst tests.*

 Tip 30 — When selecting cables, it's wise to be pessimistic about their ability to reject noise from the 220 VAC and 480 VAC power lines and noisy power supplies of a factory. CAT5 and CAT6 cables have four twisted pairs within an outer sheath to reject noise. The consistency of cable twists is vital for minimizing noise susceptibility. Good *Common Mode Rejection Ratios* (CMRR) of noise sources are the result of closely balanced connector and cable capacitance. Poor manufacturing tolerances, exposure to harsh chemicals, excessive physical abuse (e.g., bend radiuses too tight, or too much stress pulling on the cable, or running over the cable with a forklift), and even high humidity levels can seriously degrade the capacitance balance of a twisted-pair cable.

 Tip 31 — Capacitance imbalance greater than 70 pF per 100 m can introduce harmonic distortion, resulting in bit errors. The cable becomes much more susceptible to electromagnetic interference. Noise is induced more on one conductor than the other; it corrupts bits and causes transmission errors and retries. Ethernet cables vary by as much as 30 dB in CMRR.

Deviations from cable impedance over the length of the cable are common and negatively impact performance. This results in backward reflections and a condition known as return loss. *Return loss* is a summation of all the reflected signal energy coming backward toward the end where it originated. It is reported in decibels as a ratio of the transmitted vs. reflected signal.

Return loss numbers are analogous to signal-to-noise ratio: High return loss is desirable. Low return loss means smaller negative numbers in decibels; high return loss means larger negative numbers in decibels. Selecting a cable with 5% impedance mismatch instead of 15% improves return loss by up to 10 dB.

Tip 32 – Shielded Twisted Pair (STP) is naturally more noise-immune and is preferable to UTP in noisy situations. It should have at least 40 dB CMRR and less than 0.1-pF capacitance unbalance per foot.

 Tip 33 – Fiber optic is certainly more expensive but it bypasses all of these electrical issues. Especially in high-speed networks, it's a very attractive choice.

Switches Are Better than Hubs

 Tip 34 – If one section of a network is exposed to excess amounts of electrical noise, it's best to isolate that section with switches. Noise doesn't pass through switches, only packets headed for real destinations do. Hubs distribute messages indiscriminately and offer less protection against noise sources. If you must use a hub in a noisy environment, use one with some level of intelligence instead of a "buffer."

Better Cables Are Not Always Better

CAT6 cable can operate up to 1000 Mbps because of its superior bandwidth. However, this can actually cause problems in high-noise environments in a 100-Mbps network because it transmits noise more easily. Ordinary CAT5 cable is better in this situation.

Don't Skimp on Cables and Connectors

The performance difference between office-grade and high-quality cables may not make the slightest difference in your house, but it could make or break an automation system that's expected to operate reliably for many years.

 Tip 35 – The cost of cable in relation to the total cost of related equipment is quite small. If you're looking for ways to save money, this is not a place to do it. Choosing a well-designed cable will minimize your bit error rate after installation, resulting in faster throughput and fewer glitches.

Harsh Chemicals and Temperature Extremes

 Tip 36 – If your equipment is subject to washdown or exposure to corrosive chemicals, be sure to select

cables with insulation rated to withstand exposure to those chemicals, such as PUR (polyurethane). Otherwise acids, fertilizers, and petroleum can be absorbed by the cable jacket and degrade the electrical characteristics of the conductors.

Some plastics (e.g., PVC) become brittle at low temperatures, so be certain about temperature ratings. Are the cables expected to flex? Be absolutely certain you have cable that is designed for that purpose.

7.2 When You Install Cable

- If you are unable to plan the exact cable locations, add a measure of protection with armored shield or conduit.

- If physical protection or local codes necessitate using conduit, use STP wire.

- Isolate the STP shield from the conduit, since high voltages may be present on the conduit.

- Attach the STP shield to ground at only one end of the cable. Connecting at both ends creates ground loops with substantial current flow and induces noise.

- If for some reason you are required to terminate the shield at both ends, wire a Metal Oxide Varistor (MOV), a 1-Mohm resistor and 0.01- to 0.1-μF capacitor, together in parallel. This severely limits ground current except when extreme voltages are present.

- Check cables with a cable tester, not just with an ohmmeter. A tester quickly identifies continuity problems such as shorts, open wires, reversed pairs, crossed pairs, shield integrity, and miswiring of cables.

- If your cable trays are metal, they should be conductive from end to end.

Figure 29 – If you must terminate a shield at both ends, ground one side and shunt the other with this circuit.

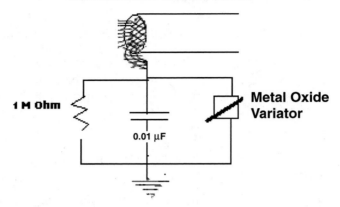

- Avoid proximity to power lines and sources of electrical transients. High-voltage lines should intersect the cable at a 90° angle.

- Maintain at least a 10-cm distance from 110 VAC, 15 cm from 220 VAC, and 20 cm from 480 VAC if you use conduit. If you don't use conduit, double those distances.

- Educate unsupervised electricians about the practices described here: Purchase a copy of this book for each of them.

7.3 How to Ensure Good Fiber-Optic Connections

Communication on fiber-optic cable is greatly affected by the cleanliness of the fiber connections, especially the cable splice and connectors. If any component is contaminated by dirt, dust, oil, etc., the transmission will be significantly degraded.

How to clean fiber-optic splices:

- Clean the fusion splice with an alcohol towel.

- Clean the connectors on S2MMs with PCB cleaner or with a cotton swab dipped in alcohol.

Fiber-Optic Distance Limits

The maximum length of a fiber-optic segment is not determined by the attenuation of the light signal but by the size of the collision domain. Exceeding allowable distances by small margins may not "crash" the network, but it will create late collisions and fragments.

These issues apply to half-duplex Ethernet with collisions. Full duplex eliminates collisions so fiber cable lengths can be greater.

Figure 30 – Popular types of fiber-optic connectors.

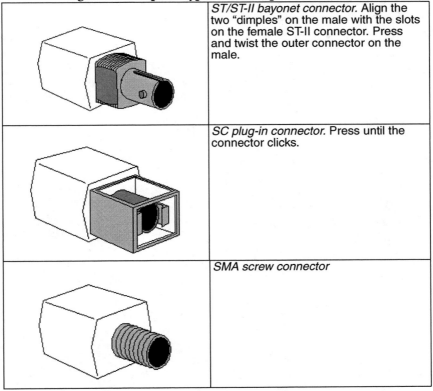

	ST/ST-II bayonet connector. Align the two "dimples" on the male with the slots on the female ST-II connector. Press and twist the outer connector on the male.
	SC plug-in connector. Press until the connector clicks.
	SMA screw connector

With full-duplex Ethernet, the multimode limit is 2 km for 10 Mbps and 100 Mbps. Attenuation causes problems for distances greater than that. With single-mode fiber, so long as the losses in the power cable do not starve the single-mode transmitter, the installation will work.

Full-Duplex Ethernet with Single-Mode Fiber

 Tip 37 – **For long distances, Ethernet needs to run in full-duplex mode. All connected segments must also be full duplex, including the switches.** Half-duplex collision domains must be connected with devices that can run the fiber link in full-duplex mode (e.g., a bridge, or using one port of a switch in full-duplex mode).

8.0—Ethernet Industrial Protocols, Fieldbuses, and Legacy Networks

As Ethernet becomes established as a major component in automation systems, it takes its place among a collection of fieldbuses and legacy networks. An in-depth investigation of all networks shows that every attribute with respect to topology, message contention, speed, and cost represents unavoidable compromises.

From a purely technical standpoint, Ethernet is not necessarily more ideal for automation applications than other networks. Even if it were, the nature of capital equipment is still such that no one is going to rip out existing equipment and wiring just because something better exists.

Reality is that (1) Ethernet must work with other network technologies, and the real world requires integration with existing networks; (2) for some applications, other networks will deliver higher performance at a lower cost.

The following questions should be considered when choosing any network:

- What is the distance requirement?

- What kind of physical cabling arrangement makes sense for this application? All of the Ethernet formats except for 10BASE2 and 10BASE5 use a star topology. This is fine for applications where devices are clustered together in groups; for others, such as a long conveyor with many nodes spaced 20 m apart, it's quite inconvenient. In this instance a trunk/drop topology (such as used by DeviceNet and CANOpen) is much better.

- What is the actual speed (response time) requirement for the most time-critical devices? Do all of the devices require that level of speed, or should some devices have a higher priority than others?

- Does the network allow you to prioritize messages?

- Do the devices you want to use support the same network standard? Are there open vs. closed architecture considerations?

- If you are developing a network-capable product, what is the hardware bill of materials and cost of software development for that network?

- How much electrical noise is present in the application and how susceptible is the cabling?

- What is the maximum required packet size for the data you are sending? If the data can be fragmented over several packets, how fast does a completed message have to arrive?

- What types of device relationships are desired (master/slave, peer-peer, broadcast)?

- Does the network need to distribute electrical power? How much current?

- What kind of fault tolerance needs to be built into the network architecture?

- What is the total estimated installed cost?

As industrial Ethernet became a hot topic, the natural next question was "How will industrial data be represented on Ethernet?" Without a standard, Ethernet will be just as confusing and proprietary as all the other networks out there.

8.1 Encapsulating Industrial Protocols in TCP/IP

In terms of carrying data, think of TCP/IP as a "flatbed truck." You can put anything on a flatbed truck you want, whether it's a bulldozer, steel coils, cartons, crates, or a mobile home. Similar, TCP/IP can be used to transport any message or file.

You can think of a traditional fieldbus as being more like a school bus, with all of the I/O data neatly arranged and placed into seats on each side of the aisle.

A school bus was designed to carry children, but a flatbed truck can carry anything—*even a school bus full of children.* And it is exactly this concept that is being used to standardize process communication on Ethernet: Putting an application layer on top of TCP/IP, just like you could transport a school bus by parking it on a flatbed truck.

Caveat Emptor

Now you're going to read about five different open standards for representing industrial data on Ethernet. There are also other proprietary standards used by some vendors. Here are some things to keep in mind:

- These protocols are not interoperable, though it is possible to define structures that make them interoperate.

- All of these protocols can theoretically exist on the same network, even though they don't interoperate.

- Yes, this is another version of the fieldbus wars, though not nearly as competitive as the last round was.

- Even within a single protocol, there are variations in the features supported.

- You should do your homework on all industrial Ethernet products you purchase, especially in terms of these protocols.

Popular Industrial Automation Protocols on Ethernet

Modbus/TCP. Modbus is the most prevalent serial protocol in automation and is most commonly transmitted on RS-232, RS-422, and RS-485. Modbus data packets can be easily encapsulated in TCP/IP.

How Modbus Works. Modbus is a fairly simple protocol. A Modbus packet consists of four parts:

Address	Function Code	Data	Checksum

When the master sends a command, the slave responds with a similarly formatted message, acknowledging receipt.

What Are "Modbus ASCII" and "Modbus RTU"? The Modbus protocol comes in two types:

- ASCII: Each 8-bit byte in a message is sent as two ASCII characters.

- RTU (Remote Terminal Unit): Each 8-bit byte in a message is sent as two 4-bit hexadecimal characters.

The main advantage of the RTU mode is that it achieves higher throughput, while the ASCII mode allows time intervals of up to 1 s to occur between characters without causing an error.

Limitations. Modbus on a *serial* network is not fast—response times of a fraction of a second are not at all uncommon. Which, of course, is why Ethernet is an attractive alternative to serial—10-, 100-, or even 1000-Mb performance is possible. Also, the simple protocol does not support complex objects and sophisticated device profiles. The master/slave orientation does not prevent peer-to-peer communication, but it requires separate "sessions" to be opened up between devices.

Why Modbus Is So Popular. Modbus is popular because it's simple, open, and because it can be used not only on copper, but on fiber-optic cable, radio transmission, and other protocols.

Another important reason is that RS-232 and -485 themselves are *not* protocols. They are transport media. Having a serial port is no guarantee of interoperability. This is still a major problem with many devices—manufacturers define their own protocols, to the chagrin of all their customers. Modbus is a great de facto standard because everyone understands it and it can be implemented on products without difficulty. Every company that makes an industrial device with a serial port should strongly consider using Modbus.

Finally, Schneider Automation (who owns the rights to Modbus and owns the trademark) has a royalty-free license, and the specifications for Modbus and Modbus/TCP protocols on its Web site are free for anyone to use. Source code examples are also available for free download to help users implement Modbus/TCP drivers.

How Modbus Is Put on Ethernet. Putting Modbus protocol on Ethernet is fairly straightforward: It is simply encapsulated in TCP/IP, and the 16-bit Modbus checksum is replaced by TCP/IP's 32-bit checksum.

Example code on how to do this is available on the Internet.

Ethernet Modbus/TCP Disadvantages. As mentioned before, Modbus is a master/slave protocol. This prevents sophisticated applications like peer-to-peer communication and some types of automatic device configuration. Ethernet does not negate the limitations of Modbus itself, although many kinds of protocols can exist simultaneously on one Ethernet network.

Figure 31 — Format of a Modbus/TCP frame.

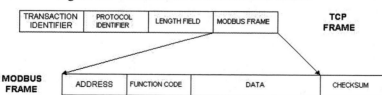

Courtesy Schneider Automation, www.modbus.org

EtherNet/IP

This is the DeviceNet/ControlNet object on Ethernet. It is backed by Open DeviceNet Vendor Association and Rockwell/Allen-Bradley and therefore has a strong political advantage. It is considerably more complex, adding expense for developers, but the complexity brings advantages.

Ethernet/IP is based on an object model called CIP (Control and Information Protocol), which is carefully mapped to both TCP/IP and UDP. It incorporates the following message hierarchies and scheduling mechanisms:

- **Exchange of basic I/O, PLC style.** All control networks have this capability: exchanging data with racks of I/O, for example, or collecting readings from a temperature controller or encoder. This is handled in a simple master/slave relationship using UDP (see Chapter 4). UDP is well suited to handling I/O control data for two reasons: UDP does not require each message to be acknowledged, so speed is maximized. And unlike TCP/IP, which is one-to-one, UDP supports one-to-many node relationships.

- **Upload/download of parameters and setpoints; transfer of programs and recipes.** In DeviceNet, these are called "explicit messages" and are sent only on a sporadic basis. A DeviceNet temperature controller uses explicit messaging to adjust or report temperature setpoints, Proportional/Integral/Derivative loop control variables, and network parameters like baud rate and node number. An Ethernet/IP temperature controller maps its data in a very similar way. Ethernet/IP uses TCP/IP for this task because this allows one-to-one communication between one device and another with full acknowledgment of a successfully received message. The extra time needed for these acknowledgments is OK because such messages are usually not as time-critical as I/O messages.

- **Polled, cyclic, and event-driven data.** Processes vary widely in the timing requirements of data exchange. *Polling* is when a master requests data from a slave on a regular basis (according to its own schedule) and the slave responds. *Cyclic* means that the slave automatically connects to the master on a predetermined schedule, such as every 100 ms. Cyclic is also called "heartbeat" messaging because it is used to tell the master it's "still alive." *Event* messaging is also called *Change of State* because the device only reports to the master when its data has changed. Ethernet/IP supports all three messaging types via UDP.

- **One-to-one, one-to-many, and broadcast.** TCP/IP is inherently one-to-one (and peer-to-peer) but this poses a significant disadvantage when many devices must be updated in a short period of time because each connection eats up valuable milliseconds. UDP allows messages to be sent to many nodes simultaneously and this saves time.

PROFInet

This is the Profibus Trade Organization's answer to the need for interoperability between automation devices and subsystems, which are linked together via Ethernet.

To understand what ProfiNet is, you must understand what it is *not*. ProfiNet is not the Profibus protocol on Ethernet in the same way that Modbus/TCP is the old familiar Modbus on Ethernet. ProfiNet is not really a "fieldbus" as the term is normally understood, either.

ProfiNet is not even Ethernet-specific; it links via TCP/IP and occupies layers 3 and above in the ISO/OSI model. Other physical layers, such as modems, WANs, VPNs, or the Internet may be employed so long as a ProfiNet device is linked to the network via TCP/IP.

An analogy to the office environment may help you understand what it is intended to do.

You work in an office with a PC at your desk, networked with a dozen other PCs and a file server. Your office LAN (Ethernet 100BASE-T) is also linked to a T1 Internet line. You open Microsoft Word and create a complex document. You write some text and create some tables. Your co-worker Jeff has a PowerPoint presentation on his PC; you open it via the network and copy and paste two different graphics images into your Word document, which are transferred intact as objects. Your other co-worker Leslie has an Excel spreadsheet, which you also open remotely and embed in your document—it's as simple as cut and paste. In this case, the Excel data is not "static," it's live: Leslie updates this spreadsheet every Tuesday, and every time you open your document, it's going to retrieve the latest data from her document on her PC. You get on the Internet and copy and paste text and graphics from a Web site, and your document also contains hyperlinks to other Web sites.

Behind this transparency among applications is a very complex object model created by Microsoft. Savvy PC users are quite accustomed to this level of sophistication and its benefits. This expectation naturally extends to the integration of business applications throughout an entire company and of course to devices in an automation system.

This is the expectation that ProfiNet was engineered to satisfy. The OLE for Process Control (OPC) software standard (see www.opcfoundation.org) was developed to create transparency between hardware devices (e.g., network and IO cards) and software applications (operator interface and programming tools); ProfiNet in fact uses components of OPC (COM and DCOM) and extends this transparency to all devices on a TCP/IP network, further defining object models for many kinds of device and programming parameters.

And rather than being specific to only one manufacturer's hardware or software (as is often the case with Microsoft), ProfiNet purports to be an industry standard available to all Profibus members.

ProfiNet is an open communications and multi-vendor engineering model. This means that a preconfigured, preprogrammed, and pretested machine such as a transport conveyor can be set up using the vendor-specific electrical devices and applications as it has been in the past.

With ProfiNet, the entire vendor-specific module (machine, electrical, and software) is represented as a vendor-independent ProfiNet component. This ProfiNet component is described within a standardized eXtensible Markup Language (XML) file that can be loaded into any ProfiNet engineering tool, and interconnections between the ProfiNet objects can be established by connecting lines from object interface to object interface.

In regards to communication and physical topology, established protocols such as TCP/IP, RPC, and DCOM are used. Data access to the ProfiNet objects is standardized via OPC. As for physical device connections, not only can devices be connected via an integrated ProfiNet interface, but also existing intelligent devices that are currently used with fieldbus networks such as Profibus can be connected to Ethernet through a gateway device called a *ProfiNet proxy server*.

Every ProfiNet object is described by an XML file that defines these parameters so that every defined data type in the system is accessible by name throughout the ProfiNet network. Integrators do not have to link devices at the bit level. It is expected that ProfiNet proxies for each different fieldbus system (Profibus, DeviceNet, Modbus, ControlNet, and others) will be developed over time, extending the transparency of large systems.

Figure 32 – The communication layers of PROFINet.

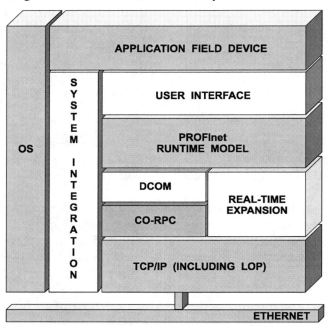

To delve into the internals of ProfiNet is beyond the scope of this book. However, more information is available at www.Profibus.com, and Profibus organization members can download the specification and source code at the site.

Foundation Fieldbus High-Speed Ethernet. This protocol uses the Foundation Fieldbus H1 process control protocol on TCP/IP.

Figure 33 – PROFINet aims to seamlessly link disparate types of control components.

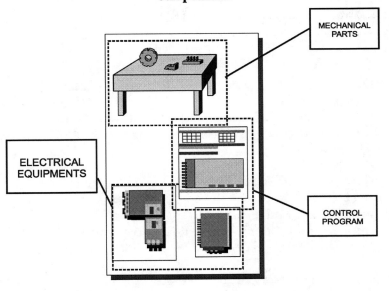

INTEGRATING DIFFERENT COMPONENTS USING PROFINET

Foundation Fieldbus H1 is a sophisticated, object-oriented protocol that operates at 31.25 Kbps on standard 4-20ma circuits. It uses multiple messaging formats and allows a controller to recognize a rich set of configuration and parameter information ("device description") from devices that have been plugged into the bus. Foundation Fieldbus even allows a device to transmit parameters relating to the estimated reliability of a particular piece of data. Foundation Fieldbus uses a scheduler to guarantee the delivery of messages, so issues of determinism and repeatability are solidly addressed. Each segment of the network contains one scheduler.

Foundation Fieldbus HSE is the same H1 protocol, but instead of 31.25 Kbps, it runs on TCP/IP at 100 Mbps. It provides the same services and transparency of network objects but operates at a higher level.

Foundation Fieldbus is specifically focused on the process-control industry and will likely be the dominant Ethernet I/O standard there. Installations in this segment of the world typically have the following characteristics:

- Very large campuses (e.g., chemical refineries) with large numbers of nodes

- Data does not have to move quickly, but there's a lot of it to move (large packets)

- Large quantities of analog data

- Hazardous area classifications such as Class I, Division 2.

Foundation Fieldbus H1 links local islands of transducers and actuators; HSE links controllers and transmits a high level of information over large distances.

IDA: The Interface for Distributed Automation. IDA was started by a group of European companies who envisioned object transparency across Ethernet networks, based entirely on open Internet protocols. IDA attempts to incorporate all aspects of a device into an open-object-based profile. The following is excerpted from IDA's April 2001 white paper:

The IDA group proposes an integrated approach for the modeling of the communication aspects and the network view of the functionality of automation devices. The IDA device communication is based on existing Ethernet communication standards and protocols (i.e. IP, UDP, TCP, HTTP, FTP, SNMP, DHCP, NTP and SMTP).

In order to realize the configuration and execution of real-time communication services IDA specifies an object-oriented model. This model is built on a hierarchy of communication objects accessible through the IDA API. The IDA API offers specific support for safety applications. An IDA communication system provides real-time and non-real-time communication services.

The real-time services are used for

- data distribution
- on-demand data exchange
- remote method invocation
- event notification

The non-real-time services are used for

- Web diagnostics and configuration (HTTP)
- file transfer (FTP)

- network management (SNMP)
- address management (BOOTP/DHCP)
- mail notification (SMTP)

The IDA real-time communication is based on the use of the Real-Time Publish/Subscribe (RTPS) protocol. RTPS uses the UDP protocol. The IDA Device object represents the complete physical station attached to the network with all its resources and components. An IDA Device has a unique device name in the automation network. It is derived from the IDA Structure. An IDA Method Server associated with this IDA Device object allow to access the methods from remote.

This concept somewhat resembles the ProfiNet model, but avoids the use of Windows standards (e.g., COM and DCOM) and uses Internet standards instead. It is also designed to specifically serve as an I/O-level solution, whereas ProfiNet is geared toward an architecture that uses a dedicated fieldbus (such as Profibus) for I/O.

Is There Interoperability Between Application Layer Protocols? Basically, no. Foundation Fieldbus HSE devices will not talk to Modbus/TCP devices, nor will EtherNet/IP devices talk to IDA devices. However, the situation is not quite as bad as it might appear.

First, they all use TCP/IP and Ethernet. So ISO layers 1–4 are already agreed upon. That's a huge step forward. Second, they all can coexist on the same wire at the same time. Third, there's nothing to prevent vendors from making devices that support more than one protocol. Some vendors do, and some devices can automatically recognize which protocol(s) are being used on the network. Furthermore, any number of protocols can be used by a device *simultaneously*. You take advantage of this on your computer every day. Finally, objects that define the relationships between protocols will be adopted by standards organizations. The issue of low-level device compatibility will slowly fade as the technology matures, and standards will compete at higher levels, that is, system-level integration of objects and profiles such as those found in ProfiNet, IDA, and OPC.

 Tip 38 – Always **be certain what protocol(s) are supported by the I/O and smart devices you purchase.**

The Ever-Popular "Embedded Web Server". It's only natural that one would ask for a Web server in an industrial Ethernet device. In theory, any smart device could be linked to a network and serve Web pages with information about its configuration or status. A Web server can easily coexist with any other protocol or application the device happens to support. But the next logical question is "what exactly will it do?"

It's best to think of a Web server as a convenient operator interface or configuration tool, an easy way to view or change settings—but certainly not a mechanism for handling high-speed real-time data between two devices.

In a temperature controller it could be used to view or change setpoints or PID (Proportional/Integral/Derivative) feedback loop coefficients. In a motor drive it might set acceleration. It's certainly convenient to do this through a Web browser. This has nothing to directly do with industrial protocols like Modbus/TCP or the real-time data that is exchanged by those protocols.

A not-so-obvious issue with putting a Web server in an embedded device is the processing power and memory required to support it. There are many factory automation devices that run on 8- and 16-bit microprocessors, such as 8051s and 80186s. It's not impossible to put a Web server on a small processor like this, but it requires a completely different mentality on the part of developers. They don't have the luxury of a PC with its powerful processor, big hard drive, and megabytes of RAM. And unlike PC-based Web servers, such as the Linux Apache Web server, the necessary software is not usually free. Fundamentally this is a cost factor that inevitably impacts the price of products in a lower-volume marketplace.

8.2 Web Services Technologies

Something is underway in the Internet world and it will not only change how the Internet works, but it is going to affect all of us in the Industrial Automation world as well. This "thing" is called Web Services and it is just beginning to have an impact on the major Internet players like Amazon, Ebay, and Google.

Web Services comprise all the communications systems, protocols, and practices for exchanging data in a well-understood format between any two computer systems. More than that it includes the ability for a "computer" to search for the information, extract it from its source, format it for presentation, and deliver it to a requestor. This would all be done using open protocols, standard data exchange formats, and common directories of information.

Lest you think that this is science fiction, these Web Services are exactly what Amazon, Ebay, and Google have been working on for a couple of years now. Amazon has published a Web Services API and SDK (Software Development Kit) for electronically accessing its database of products. This information can be electronically extracted, formatted for presentation in any format on any device at the discretion of the requestor. Amazon partners are using this technology to present real-time reviews, stock levels, prices and information from the Amazon database to customers using PDAs, TVs, telephones, and assorted other devices. Ebay, Google, and others all have similar programs for their products.

Several technologies make this happen but before we get to them, we should talk a little about HTML. You probably know the term but may not know what it means. HTML is the language that describes how to format a Web page. It is an excellent tool for creating forms with graphics, pushbuttons, checkboxes, hyperlinks, and all the other features we use everyday on the Internet. If you've ever clicked on your browser's View menu and selected "Source" from the drop-down list, a Notepad window opens up containing the raw HTML language of your Web page.

At first you might note that what pops up is all text, as opposed to binary data. It is human readable. You may also notice that it is highly structured. Brackets and codes surround the words that appear on the page. This standard format is used to display Web pages around the globe every day and it is the true power behind human exploration of the Internet. If everyone's Web page used a unique structure, the Internet would be a Tower of Babel with no ability to pass information to anyone.

While HTML is fine for human viewing and interaction with Web pages, it is extremely weak when it comes to other forms of communication. For example, HTML doesn't work well on a cell phone. It doesn't translate to audio delivery very well and is not structured right for PDAs, Notepads, and machine-to-machine communications. In machine-to-machine communications, the ability to change the data delivered as circumstances change is very important. Static HTML pages are completely inadequate in this area.

XML (eXtensible Markup Language)

XML is the language of choice for transferring data between two computers whether they are a PC, a temperature controller, a programmable controller, or an industrial drive. XML has the unique ability to transfer data and data descriptions in a format that is hardware independent, software independent, extensible and readable by any other system. The following factory floor example helps illustrate some of the shortcomings we face with the technologies we've used for the last twenty years.

Imagine you have a weigh scale at the end of your production line. The weight delivered by this scale is important for your control system as it needs to reject under and overweight packages. It is important to your logistics systems as they must load that package into a truck, and it is important to your production control computer, which needs to record how much is produced this hour, day, week and month. There are a lot of uses for this data.

As is typical for this type of system, the weigh scale delivers a string of bytes in some format every time it completes a weighing operation. A typical string of bytes is shown in Figure 34.

Figure 34 — Traditional Weigh Scale Data

Byte 1	Byte 2	Byte 3	Byte 4	Byte 5
Header Data	Weight in lbs	Weight in Gm	I/O Bits	Trailer Data

The header and trailer data let you identify the start and end of the packet but you still have some questions. How exactly are Bytes 2 and 3 formatted? Do these bytes have the actual weight or are they scaled? Are they in some kind of floating-point format? Once you learn more about the data format you can program your control system and all the other systems to receive this data.

Now if the scale maker revises their scale or you buy a new scale you are faced with repeating this effort. The new scale data now looks like Figure 35.

Figure 35 — Weight Scale Data Revision 2.0

Byte 1	Byte 2	Byte 3	Byte 4	Byte 5
Header Data	Weight in lbs	Weight in lbs	Weight in Gm	Weight in Gm

Byte 6	Byte 7
I/O Bits	Trailer Data

They've expanded the data in the packet and now all the systems dependent on this data are unable to access it.

If you expand this example to the hundreds of factory-floor devices each containing their own data encoded in their own special format, you get an idea of the problem XML is designed to solve. Using XML, a device transmits a text data file that not only contains the data but carries descriptions of the data as well. Here is what an equivalent XML file might look like:

Figure 36 — Weigh Scale Data Using XML.

```
<?xml version="1.0" encoding="ISO-8859-1" ?>
-_<Scale_Data>
    <from>Company XYZ Scales</from>
    <SrcAddr>15</SrcAddr>
    <LbsWeight>152.751</LbsWeight>
    <GmsWeight>69286.69</GmsWeight>
    <LbsWeight100>1.52751</LbsWeight100>
    <GmsWeight100>692.8669</GmsWeight100>
    <IOPoints>00001011</IOPoints>
</Scale_Data>
```

The XML data file is superior to the traditional data packet in a number of different ways:

1. The XML data file is completely hardware independent. Where values expressed in binary data formats are dependent on the byte order of the computer, values expressed in text are independent and transportable to any computer system.

2. The XML data file is completely software independent. Any IT program, PLC program, Basic program or the like can read it, find the appropriate tags and extract the data. Macros could be written for a spreadsheet or word processor to process XML data.

3. An XML file is human readable and can be directly displayed by most current browser editions. As a human-readable format, the file is easy to troubleshoot.

4. Data in XML files is extremely extensible. New data can be added at any time. As long as the current descriptors or "tags" as they are know are not changed the receiving software can continue to extract the data it needs no matter how much the rest of the data file has changed.

5. XML data is fixed format. All data is ASCII format and can be easily manipulated into the native binary format of the receiver.

The syntax of XML files is very straightforward and clear. The first line is a declaration describing the XML version number, which validates the file. The next line is the root element. The root is the parent element for all the other elements in the file. You will notice that every element including the root has a start tag of the form "<tagname>" and an end tag of the form "</tagname>". Data is contained between the start and end tags.

XML is not good for hi-speed, dedicated control applications. These applications require a small number of bytes delivered in real time. XML applications are applications where data must be delivered to widely different and disparate and sometimes, unknown destinations in a common, well-understood format.

SOAP (Simple Object Access Protocol)

XML provides a way to "broadcast" data in a well-understood way but it doesn't provide the structure necessary to request data from another application and get that data. SOAP provides that functionality. SOAP is a special case of XML. A SOAP message is an XML message with special SOAP elements. In fact, the "root" of a SOAP XML file is a SOAP Envelope element.

SOAP is an applications layer communications protocol. It provides an environment where one application can request data from another application in a well-understood, hardware and software independent manner. The two applications may be in the same computer running under the same operating system or may be across the globe running widely different operating systems using completely different hardware.

SOAP messages consist of three parts: an envelope that defines a framework for describing the message and how it can be processed; the encoding rules for expressing data types; and a convention for representing remote procedure calls and responses.

One of the most important features of SOAP is that it is designed to work under HTTP. This provides tremendous power for Web pages to request data from any server.

SOAP has the potential for becoming a very important protocol for delivering factory floor data to servers inside and outside an organization.

9.0–Basic Precautions for Network Security

The subject of protecting your data from hackers and viruses, corporate espionage, and cyber-terrorism with firewalls, security keys, passwords, and their associated organizational procedures easily occupies an entire book rack at a technical bookstore. A quick search on "network security" at www.amazon.com turned up 203 books. No book called *Industrial Ethernet* can possibly do justice to this subject. The best we can do is highlight some key concerns that you should think about. From there, you can research firewalls, encryption schemes, and Virtual Private Networks to your heart's content.

The first thing to remember is this: The most probable cause of problems is not the environmental extremist who sees your smokestack on his way to work and decides to launch a virtual terrorist attack on your factory. It's more likely to be related to the ubiquity of PCs with Ethernet cards, the ease with which your own employees can "hang stuff on the network," and careless or nonexistent *internal* security measures. Accidental problems are more common than deliberate ones. But you should be prepared for both.

The following guidelines will help you guard against the most common problems:

- **Never mix your office LAN with your industrial-control LAN.** They should be separated by a firewall, or at minimum, a bridge or router. That firewall also serves as a convenient boundary between the loyal, dedicated, competent automation engineer, and the egotistical control freak from the IT department whose mission in life is to discredit the engineering department and take over the planet. A control network and a business LAN have two entirely different purposes and their interaction should be closely controlled.

- **Industrial Ethernet needs to be viewed in at least two categories: a control-level industrial Ethernet and an I/O-level industrial Ethernet.** This means each manufacturing cell will have its own Ethernet network, possibly more than one. Ideally those networks will be isolated as well.

- At the control level, prioritization and security can be easily overlooked. **The most common instances of industrial sites being "hacked" are the result of well-intentioned employees, not outsiders.**

- Another hazard is connecting consumer "plug and play" devices to your factory LAN. A printer, for example, might flood the network with traffic with a "broadcast storm" as it tries to self-configure or advertise its presence to all nodes on the network.

- **Faulty devices, for example defective NIC cards, can vomit zillions of bad packets** (i.e., *runts,* which are abnormally short Ethernet frames) **into your network.** Using switches instead of hubs limits the effect of such problems. Diagnostic tools can locate the source of bad traffic.

- **Duplicate IP addresses can deactivate devices that otherwise appear to be perfectly functional.** This is especially common when replacing devices, and is a very perplexing problem to trace.

- **Passwords often stay the same for years, and are often easy to guess.**

- "Routing Switches" or "Level 3 Switches" can logically divide networks on the basis of IP address, IP subnet, protocol, port number, or application, completely blocking traffic that does not fit a precise profile. This offers substantial protection against broadcast storms and faulty packets while allowing specific data to freely pass between the business LAN and the factory network.

- **It's unwise to assume that your industrial Ethernet products themselves have any security features at all.** You should minimally use inspection-type firewalls (such as packet filters) to control access that is based on a combination of IP source address, destination address, and port number. This is by no means completely hacker-proof, but it should keep the well-meaning employees out.

10.0—Power over Ethernet (PoE)

Power over Ethernet technology allows Ethernet devices to receive power and data over their existing LAN cabling without modifications to any of the existing Ethernet infrastructure. PoE radically reduces installation costs by eliminating the conduit, power wires, and installation labor required to install an Ethernet device.

10.1 What is PoE?

Everyone who has ever used a standard phone line is familiar with a network powered device. Simply plug the phone into the jack and you have a network connection. Power over Ethernet, IEEE standard 802.3af, performs the same function for Ethernet devices. In fact, PoE will really enable the widespread use of IP telephones.

PoE brings the ability to connect and power devices like barcode readers, RFID Systems, wireless access points, and Web-based security cameras. PoE enables a whole new generation of networked devices. Because there is no need for the device to be anywhere near a wall socket, we can expect a plethora of innovative applications, from vending and gaming machines to building access systems and retail point-of-sale systems.

There are two types of PoE devices: Power Sourcing Equipment (PSE) and Powered Device (PD). PSE devices supply network power to PD devices. Before supplying network power a PSE device must first determine if the end device is PoE-enabled. Power is never delivered to a non-PoE-enabled device.

10.2 What Pins Are Used on the CAT5 Cable?

A standard CAT5 Ethernet cable has four twisted-wire pairs. Only two of these pairs are required for 10BASE-T and 100BASE-TX operation.

The PoE specification provides two ways for a PSE device to deliver power:

- Option 1: Power is supplied on the two spare pairs. Pins 4 and 5 are tied together to form the positive supply while pins 7 and 8 are tied together to form the negative supply.

- Option 2: Power is supplied on the data pairs, the same lines carrying your Ethernet messages. Since Ethernet pairs are transformer coupled at each end, it is possible to apply DC power to the center tap of the isolation transformer without affecting the data transfer. In this mode of operation, the pair on pins 3 and 6 and the pair on pins 1 and 2 can be of either polarity.

All PSE devices are required to support both power options. A PD device can support either option.

10.3 How Much Current Is Supplied?

A lot of work was done to determine how much voltage could be applied to an Ethernet device. There were several constraints on the system design:

1. PoE devices must not present a safety hazard. The voltage must be low enough to preclude an electrical shock.

2. No changes to existing cable.

3. No interference to the network data transfer.

4. No damage to non-PoE-enabled devices.

5. No requirement for extensive testing or certification of PoE devices.

After much analysis the IEEE standard committee picked a 48-VDC system that could deliver up to 12 W of power to a PoE-enabled device. At this voltage, common electrical standards don't require certification and testing, CAT5 cable is an acceptable delivery medium, and there is enough power to provide some decent functionality in the end device. At this voltage, a PD device can count on 350 mA at 37 V (12.95 W) after cable and other power losses are deducted. This number is a function of the limitations of CAT5 cable and the lack of a requirement for stringent device testing below 57 V.

To prevent power delivery to non-PoE-enabled devices, the standard includes a mechanism that a PSE can use to interrogate a device to determine if it is PoE enabled. This "discovery" mode allows a PoE to maintain compatibility with existing, non-PoE-enabled, equipment.

In "discovery mode" a PSE applies a small current-limited voltage to an end device. The PSE is looking for the presence of a 25-Kohm resistor in the remote device. If it finds the resistor, power is supplied. If not, the PSE marks the port as non-PoE-enabled and power is not applied to that device. The PSE continuously repeats the test in case a new device is connected on that port.

Once a port is determined to be PoE enabled and power is applied to it, it must continue to draw a minimum current. If it doesn't (device unplugged), the PSE removes network power and marks the port as non-PoE-enabled. The discovery process is then repeated until a PoE-enabled device is found.

10.4 What Are the Advantages to PoE?

PoE technology is driven by the need to install wireless access points in locations without AC power at a cost that isn't prohibitive. Depending on the location of the access point, the cost of conduit, wire, and labor to deliver power to an access point was sometimes many times more than the cost of the access point hardware.

Even though wireless access points were the driving factors behind PoE, it now includes applications in many other areas including:

- Remote security cameras
- Data-entry terminals
- Remote displays
- Vending machines
- Gaming machines
- Point of sale terminals
- Thermostats
- Voice over Internet Protocol (VoIP)

Another huge potential application for PoE is the charging of battery driven devices like cell phones, pagers, PDAs, and the like. These devices now all have an assortment of special cables and adapters. If you travel overseas, you sometimes need an adapter for every country you visit as the power line connector can vary from country to country. If your battery powered devices were equipped with standard RJ-45 connections, every Ethernet connection (standard over the world) could potentially become a source for charging your device.

PoE also provides much needed flexibility to system managers. Built into PSE devices is the ability to manage the power delivery to PDs remotely. Using Simple Network Management Protocol (SNMP) functionality a device can be powered on or off remotely allowing a device to be rebooted from anywhere in the world.

10.5 How Do I Get Started with PoE?

There are two ways to begin using PoE. First, you could replace your existing switches with PoE-enabled switches, also called "inline power" devices. The PoE-enabled switches include a power supply connection that provides the source of the network power.

Alternatively you can begin by using a mini-span hub, also know as a "power hub". This device is inserted between your switch and the network devices. It provides PoE functionality for non-PoE-enabled devices. You simply disconnect your device from the switch and plug it into the hub and then connect the hub to the port on the switch. While a mini-span hub is a way to maintain your existing infrastructure, it has the disadvantage of adding to the wiring, power, and space requirements of your wiring closet.

Prior to starting work with PoE, surveying your existing wiring is mandatory. Are all your devices serviced by CAT5 cable? If not, you will want to upgrade your cable to CAT5. Second, is all your CAT5 wired properly? Many times, installers knowing that there were unused wire pairs in CAT5 cable will "double up" devices on a single cable. PoE will not work with non-standard wiring.

You will also want to consider the need for a UPS system to backup the power delivered to your network. If you are providing critical services to your company, a UPS is probably another addition you will want to make to your wiring cabinet.

Finally, check the version of your SNMP management software. Make sure that your version supports PoE.

10.6 Resources

* http://www.iol.unh.edu/consortiums/poe/ - Power over Ethernet Consortium website

* http://www.ietf.org –Internet Engineering Task Force (IETF) for documentation on Internet Drafts

* http://www.ieee802.org/3/af - IEEE Power over Ethernet Web pages

11.0—Wireless Ethernet

Public wireless LANs are showing up everywhere. You can now find Web access in Starbucks, McDonalds, airports, convention centers and hotels. Thousands of companies are placing wireless systems in warehouses, retail stores, conference rooms and, in some cases, throughout their facilities. The radio waves that transmit data in these applications make this technology different than wired Ethernet networks. Although the contents of the base Ethernet packet is the same as a wired packet, the terms, technology, procedures and practices for operating a wireless Ethernet system are very different.

11.1 A "Very" Short Technology Primer

IEEE standard 802.11 defines the wireless communication method used in today's wireless enterprise networks. Unfortunately, there is an alphabet soup of standards within this specification for us to sort through. Before 802.11 Part a (802.11a for short) was developed, a version with less throughput, 802.11b, was adopted by some companies. After those standards, there was 802.15.1, also known as Bluetooth, the security standard 802.11i, and others like 802.11g. Fortunately, most of this headache really belongs to the wireless equipment manufacturers. All we need to understand is the basic differences between the two standards most commonly used today, 802.11a and 802.11b. These two standards are summarized in Table 11-1.

Table 11-1 – Comparison of 802.11 Part A and B

802.11a	802.11b
54 Mbps in 5-GHz Band	11 Mbps in 2.4-GHz Band
Shorter Distances	More Commercial Applications
Higher Throughput	Frequency Conflicts with common devices like Microwave Ovens
Generally Less Interference	Can be lower cost than 802.11a
Less Crowded Frequency Spectrum	Adequate throughput for email, Web browsing

802.11a is generally applied to higher throughput applications like video, high bandwidth data, and some audio. It is also more noise immune and can more easily accommodate large numbers of users, which may bog down an 802.11b network. Of course, the greater speed, functionality, and noise immunity of 802.11a are obtained at a higher price point.

Both 802.11a and b are grouped in the wireless network domain as WiFi networks. These networks span from 10 to 100 m. In addition to WiFi, some people talk about Wireless Personal Area Networks (WPAN). These networks are for desktop applications like printer sharing, telephone handsets, and the like. WPAN networks usually work within the 0 to 10 m range. Large networks are known as Wide Area Networks (WAN) and are used to link buildings together. In this chapter, we will focus on WiFi, the most common networks for industrial applications.

When thinking about wireless communications, think about air as just another communications medium. With wired Ethernet, we use CAT5 cable to physically transfer data from one point to another. As we discussed in earlier chapters, we have some physical interface to the media and a software interface known as a Media Access Controller (MAC). When we look at wireless, we have the same components. We have a medium, a physical interface, and some media access control. The medium is air. The Physical Interface is the hardware that converts the bits and bytes of the Ethernet message to a wireless signal. The MAC is the software/hardware layer that monitors the channel and controls access, monitors the channel for collisions and schedules messages for transmission.

A few basic physical communications standards used in wireless communications include spread spectrum, OFDM, and infrared. Of these the spread spectrum standard is the most interesting. In spread spectrum, data is transmitted in using a number of channels. A part of the message is sent on one frequency, the device "hops" to the next channel, and then sends the next part of the message. The number of hops, the channels used, and the sequence account for some of the differences in the different wireless networks. In general, the more hops, the less opportunity there is for your entire message to get destroyed by outside interference.

No matter which physical standard is used, interference can disrupt the operation of your wireless system. Interference sources can include Bluetooth devices, cordless phones, and neighboring wireless LANs. As the number of wireless devices continue to grow, increasing attention needs to be made to these interference sources <u>before</u> your wireless system is deployed.

Another source of lost packets and lower bandwidth is attenuation. Attenuation is a decrease in the amplitude of the signal as it radiates out from the source. Signals are attenuated by everyday objects including walls, machinery, pillars, floors, and furniture. The range of any wireless LAN is a function of the number and type of objects between the transmitter and receiver. The completion of a site survey prior to deployment identifies the presence of attenuators and locates access points to obtain the greatest coverage through the facility.

Several vendors make site survey tools to assist vendors in the site survey process. These tools are designed to find sources of interference and attenuation so that access points can be located or aimed to get the greatest throughput.

11.2 Access Points

The typical low-cost access points available from your local retail computer store provide a shared wireless network where clients take turns transmitting their data. These access points are similar to an Ethernet hub in a traditional wired Ethernet network. Just like their wired cousin, a wireless LAN wants only one message at a time on the network. However, just like the wired version, the more users on the network, the more collisions, the more packet retransmissions, and the less throughput for everyone.

The more expensive wireless access points are more like Ethernet switches than hubs. In a wired switch, each node can transmit at any time and the switch resolves a lot of the collisions by retransmitting the message only to the selected destination not to everyone on the switch. Since everyone can transmit simultaneously, there are fewer collisions and greater throughput. Many of these wireless switches attain the same type of simultaneous transmission and greater throughput by using directional antennas to aim signals at a particular node and using multiple channels. Besides simultaneous transmission and fewer collisions, these types of access points provide the network with much better range. Again, these wireless switches achieve this greater bandwidth at a significantly higher cost.

11.3 Mesh Networks

An alternative to expensive switched access points is a wireless mesh network. Mesh networks are networks of wireless nodes that can relay messages around the mesh to locate a specific node. These networks can easily accommodate sudden interference, are more mobile, and are more adaptable to a changing environment. Mesh networks can transfer messages in any number of relays through a varying node sequence. If a message is typically transferred from Node A to Node D through Nodes B and C, it will automatically transfer a message between A and D using Nodes B, G, M, and T if Node B suddenly loses access to node C.

The downside to Mesh networks is the lack of common standards. Most Mesh network systems are more proprietary than open with routing software particular to their brand of mesh network. Users should think long and hard before committing their future to a single source system.

11.4 Security

Wireless security is such an important and complex topic that it really warrants an entire book. Instead, this section will attempt to provide a basic explanation of the problem and how to deal with it.

Because radio waves propagate through walls and outside your physical space, wireless LANs pose a threat to the integrity of your IT system and the operation of your entire enterprise. For example, the transmissions of a wireless LAN can be passively monitored for a long distance from your facility using directive antennas. If you don't implement the minimum standard security mechanisms built into your wireless devices, outside eavesdroppers can read emails and access files sent between users.

There are any number of methods individuals may use your wireless LAN to harm your facility. Some of the most common include:

- SSID sniffing
- WEP encryption key recovery attacks
- ARP Poisoning
- MAC Spoofing
- Access Point Password attacks
- Wireless end user station attacks
- Rogue AP Attacks

- DOS (Denial of Service) Attacks
- Planned Cordless phone Interference

The most common method to prevent these attacks and others like them is to activate the Wired Equivalent Privacy (WEP) security included in most wireless devices. WEP encrypts the body of each data frame and is designed to prevent unauthorized users from detecting email addresses, user names, passwords, and viewing sensitive documents. Unfortunately, weaknesses in the WEP security system have been detected and hackers can sometimes break into WEP-protected systems in as little as 24 hours.

One of the weaknesses of the current WEP system is the encryption keys used by the wireless transmitter and receiver. Both ends must know the encryption key. But because management of these keys is such a headache for the wireless systems administrators, these keys are hardly ever changed. Without frequent re-keying of a wireless system, hackers can get months and months to work at discovering the keys and cracking the WEP security.

The IEEE 802.11 Working Group is addressing the weaknesses of WEP by creating a new standard, the IEEE 802.11i standard, which include not only extremely reliable encryption of data packets but also the rotation of encryption keys. The key rotation feature alone will go a long way toward discouraging all but the most determined hackers. Until 802.11i is ratified and accepted the WiFi Alliance has adopted WiFi Protected Access (WPA). This standard is very similar to 802.11i and enables vendors to begin integrating much better security into their wireless devices.

Another security consideration that is often neglected is wireless access to your enterprise network over public wireless systems. Even though WPA and 802.11i deal with the wireless security your data must still be transferred over a wired Ethernet system from the access point outside your facility. To protect your data over the wired systems, companies should ensure that users have virtual private network (VPN) client software. A VPN encrypts data all the way from your remote user to the corporate VPN server.

11.5 The Advantages

The advantages of wireless LANs are numerous. There are, of course, economic factors. Components such as Access Points, wireless adapters and other network hardware are continually dropping in price. Installation is faster and less expensive without the requirement to run cabling and install a wall out for each user. Wireless access is faster as the system is immediately operable once the Access Point is configured. And if your Access Point uses Power over Ethernet (PoE), no power cables are required.

Over the long term, the ability to reconfigure your office without reconfiguration of your network, the lack of cabling to be mistakenly cut and the higher reliability of wireless LANs are significant advantages of a wireless system. Some argue that the long-term cost of ownership of a wireless LAN is less than its wireless counterpart. They argue that not only is the overall per point cost lower but the productivity advantages of a wireless system over a wired system are significant.

Index